Barrier Islands

of the **Florida Gulf Coast Penninsula**

The Most Complicated
Barrier Island System
in the World

Richard A. Davis Jr.

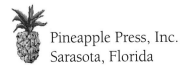

Pineapple Press, Inc.
Sarasota, Florida

Library of Congress Cataloging-in-Publication Data

Names: Davis, Richard A., Jr., 1937-

Title: Barrier islands of the Florida Gulf Coast peninsula / Richard A. Davis Jr.

Description: First edition. | Sarasota, Florida : Pineapple Press, 2016. |
  Includes bibliographical references.

Identifiers: LCCN 2015028265 | ISBN 9781561648085 (pbk. : alk. paper)

Subjects: LCSH: Barrier islands--Florida--Gulf Coast. |
  Geology—Florida—Gulf Coast. | Barrier island ecology—Florida—Gulf Coast. |
  Gulf Coast (Fla.)—Geography. | Gulf Coast (Fla.)—Environmental conditions.

Classification: LCC F317.G8 D38 2016 | DDC 975.9--dc23

LC record available at http://lccn.loc.gov/2015028265

Inquiries should be addressed to:
Pineapple Press, Inc.
P.O. Box 3889
Sarasota, Florida 34230
www.pineapplepress.com

Design: Doris Halle
Printed in the United States

Contents

Contents continued

Preface

Valuable real estate. Beach-lover's paradise. Marine life nursery. Mainland protector. Accumulation of mobile sand. All of these terms accurately describe a barrier island. Barrier islands are those long, narrow strips of land that separate the ocean from the mainland with open water in between.

For real estate investors and developers, the value of barrier islands is in their nearly continuous beaches—or, more precisely, in their appeal to beach-loving vacationers and second-home owners, retirees, and the commercial interests that are associated with those lifestyles. The beaches of many barrier islands are vital nesting habitat for sea turtles and home to ghost crabs, sand hoppers, and other creatures, while the landward sides, often covered by wetlands, serve as nursery grounds for many species of shellfish and finfish as well as rookeries for shorebirds.

Coastal areas are always subject to the actions of wind and water, and when those forces are driven by storms, the consequences can be catastrophic. While barrier islands cannot completely protect the mainland from the devastation of hurricanes and the like, they can help temper the effects by absorbing some of the storm energy and dissipating the force of wind and water. Reduced to the most basic terms, barrier islands are the product of the movement and accumulation of sand by ocean waves and tides, which means they are constantly changing, separating, combining, and even disappearing over time, sometimes slowly over centuries and decades, and sometimes in only days.

Barrier islands and barrier spits are present all over the globe, from the Equator north to the Arctic and south to the Antarctic, accounting for some 15% of Earth's coastline. In the United States, they are present to some extent on the Pacific coast, while nearly the entire lengths of the Atlantic and Gulf of Mexico coasts are lined by barrier islands. The side of the Florida peninsula that borders the Gulf of Mexico is considered to have the most diverse and interesting barrier island system in the system in the world, and that is the primary topic of this book.

What follows will be of interest to anyone who lives on or near or visits a barrier island, as well as anyone who is interested in barrier islands in general and those along Florida's Gulf coast in particular. The book consolidates information about what scientists understand about the processes that create and affect barrier islands as well as the various types of barriers and their characteristics. For the layperson, concepts and terminology that appear in italics are explained and defined in the comprehensive glossary at the back of the book.

Florida Gulf Pennisula Barrier Inlet System

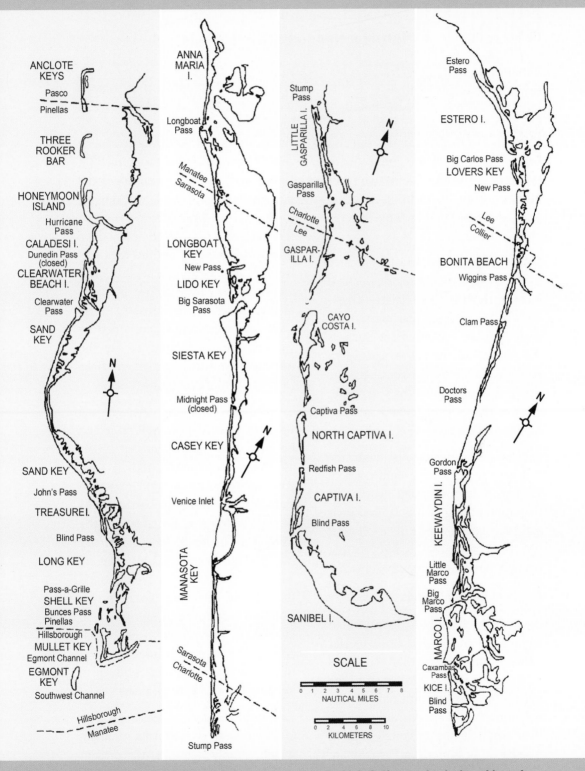

ANCLOTE
KEYS

Pasco
Pinellas

THREE
ROOKER
BAR

HONEYMOON
ISLAND

Hurricane
Pass

CALADESI I.
Dunedin Pass
(closed)

CLEARWATER
BEACH I.

Clearwater
Pass

SAND
KEY

N

SAND KEY

John's Pass

TREASURE I.

Blind Pass

LONG KEY

Pass-a-Grille
SHELL KEY
Bunces Pass
Pinellas

Hillsborough

MULLET KEY
Egmont Channel

EGMONT
KEY

Southwest Channel

Hillsborough
Manatee

ANNA
MARIA
I.

Longboat
Pass

Manatee
Sarasota

LONGBOAT
KEY

New Pass
LIDO KEY

Big Sarasota
Pass

SIESTA KEY

Midnight Pass
(closed)

CASEY KEY

Venice Inlet

N

MANASOTA
KEY

Sarasota
Charlotte

Stump Pass

Stump
Pass

LITTLE GASPARILLA I.

Gasparilla
Pass

Charlotte
Lee

GASPAR-
ILLA I.

CAYO
COSTA I.

Captiva Pass

NORTH CAPTIVA I.

Redfish Pass

CAPTIVA I.

Blind Pass

SANIBEL I.

N

SCALE

0 1 2 3 4 5 6 7 8
NAUTICAL MILES

0 2 4 6 8 10
KILOMETERS

Estero
Pass

ESTERO I.

Big Carlos Pass
LOVERS KEY

New Pass

Lee
Collier

BONITA BEACH

Wiggins Pass

Clam Pass

Doctors
Pass

N

Gordon
Pass

KEEWAYDIN I.

Little
Marco
Pass

Big
Marco
Pass

MARCO I.

Caxambas
Pass

KICE I.

Blind
Pass

This segmented map shows the barrier island system along Florida's Gulf coast that is the subject of this book. The Anclote Keys, top left, are the northernmost islands in the system, and Kice Island, bottom right, is the southernmost. Broken lines show county boundaries.

Chapter 1 — Introduction to Barrier Islands

For the purposes of this book, the topic of barriers includes not only true islands—i.e., those that are not connected to any land mass—but also barrier spits, which are connected to the coast at one end. These two types of barriers are the same in all other ways; they contain the same specific environments.

Barrier islands are most extensive in tectonically stable regions. These include trailing edge coasts—coastlines of land masses that are moving away from an oceanic ridge, such as the eastern side of North and South America; and *mediterraneans*—marine basins that are smaller than an ocean, such as the Gulf of Mexico and the Mediterranean Sea.

Before considering the specific barrier complex of the Gulf coast of Florida, it is appropriate to discuss the origin of barrier islands and their *morphodynamics*. This will be followed by the general causes and effects that operate in these complex systems.

Origin of Barrier Islands

There is still some disagreement among experts about how barrier islands form and whether they form in multiple ways. It was not until the middle of the nineteenth century that a serious theory was formulated for their origin. It was presented in 1842 by Elie de Beaumont, a Frenchman.

He postulated that barrier islands were formed by the emergence of sand accumulations parallel to the shore, the result of upward *shoaling* caused by wave action. Later, in 1885, Grove Karl Gilbert, studying Lake Bonneville in Utah while working for the U.S. Geological Survey, theorized that barriers were formed by *longshore transport* of sand. The third idea was from William D. McGee in 1890, who cited the drowning of beach ridges as the source of the barriers. These three concepts were championed by various researchers until late in the twentieth century. Some even chose to support more than one theory.

The upward-shoaling idea of de Beaumont was not well received because most people viewed waves as being destructive rather than constructive. The idea of longshore transport to develop barrier spits was considered reasonable for the formation of that type of barrier but not for barrier islands. Several workers in the 1960s championed the idea of drowning beach ridges to be a reasonable concept. In actual fact, there are problems with that concept because it requires sea level to change very rapidly while also keeping the beach ridge intact.

One of the primary axioms in geology is the concept of uniformitarianism, which, simply stated, is "the present is the key to the past." If we study barrier islands on the coast over a period of time, we should be able to observe the formation of one or more of these barriers and from those observations we can determine what caused them to form. As it turns out, the author was one of the first people to do this and it was done on the Florida Gulf barrier island complex that is the subject of this book.

Another possible explanation of the origin of the barriers that we see on the coast today is that they formed on the continental shelf and migrated landward to their present position. The barriers still had to have formed by one of the methods described above. There are linear sand bodies offshore on the continental shelf that probably were barriers. Rapid rise in sea level would have "drowned" them

and left them behind. Destruction of barriers is more likely than migration based on evidence from existing barriers.

The barrier island that we can use to demonstrate this concept is now known as Shell Key (formerly called North Bunces Key), located in Pinellas County, Florida, just north of the entrance to Tampa Bay (figure 1.1). This sand body first became *supratidal* (remaining above water at high tide) in 1962. The continuing development of the island has been recorded to the present time. The origin is a matter of abundant sediment and continued low wave height that permits the concentration of sand. Similar development of barrier islands has taken place elsewhere in the world in the twentieth century and is continuing to the present.

Figure 1.1

Recent aerial photograph of North Bunces Key (Shell Key)

There is no documented example of a barrier island forming through the drowning of a beach ridge, contradicting McGee's theory. More than one barrier island has developed on the Florida Gulf peninsula from upward shoaling, in keeping with de Beaumont's theory. The keys to barrier island formation are small waves and abundant sediment plus a place for the sediment to accumulate.

Barrier Island Types

The various nearshore processes coupled with the available sand combine to produce a wide range of shapes and sizes of barrier islands along Florida's Gulf coast. The *tidal range* of this barrier coast is less than one meter at spring tide, while the tidal range on the non-barrier *tide-dominated* coast to the north and south is greater. Barrier islands are limited to what are called *wave-dominated*

or *mixed-energy* coasts. Barriers require significant wave influence to form. This produces shore-parallel sediment accumulations, whereas tide-dominated coasts display tidal creeks that have shore-normal orientations (figure 1.2). Significant changes can take place even though the waves and tidal range are both small. The important condition is the energy relationship between waves and tides. This is explained below.

The type of barrier island that evolves from the original sand shoals is not important for determining how the barriers will react to sea-level rise. It is only somewhat important for human development. The mixed-energy barriers are better for development activities than wave-dominated barriers simply because they present more surface area. As experience has shown, all types of

Figure 1.2

Oblique infrared photo of the tide-dominated marsh coast in the Big Bend, north of the barrier island system

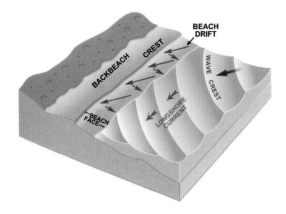

Figure 1.3

Diagram showing the refraction of waves as they approach the shoreline. This generates currents called longshore currents that move parallel to the beach.

barrier islands have been completely developed with all types of structures and all locations along them.

Wave-Dominated Barrier Islands

These are long and narrow islands where the combination of waves and *longshore currents* accumulate sediment to produce a supratidal sand accumulation. When the waves approach the shoreline of this sand accumulation in shallow water at an angle, a longshore current is produced (figure 1.3). This causes sand to move along the shoreline and extend the island. Most commonly this extension is in a dominant direction for a given island, but there are some islands where this takes place in both directions. On the Gulf coast of Florida this happens on Anclote Key (figure 1.4), Three-

Figure 1.4

Oblique aerial photo of Anclote Key, an example of a wave-dominated barrier island and the northernmost barrier on this coast

Figure 1.5

Aerial photo showing the widespread mangrove community on the landward side of Lover's Key

Figure 1.6

Photo of Three-Rooker Bar just south of Anclote Key showing washover fans that extend into the backbarrier bay

Rooker Bar, and Sand Key, examples of barrier islands that include a wide range in lengths.

Wave-dominated barrier islands on Florida's Gulf coast tend to be narrow, with wetlands on the landward or backbarrier side (figure 1.5). These wetlands are comprised of mangrove *mangals* and salt marshes. Because this coast has a limited sediment supply offshore, the islands tend to be low in elevation with only small sand dunes, if any at all. As a result the islands are washed over during even

Figure 1.7

Willy's Cut on Caladesi Island, a) as viewed from the air, and b) across the channel cut by the storm surge associated with Hurricane Elena in 1984

modest tropical storms. Washover processes transport considerable sediment to form what are called *washover fans* (figure 1.6).

The narrow and low island is quite vulnerable to breaching by waves and currents during *storm surge* conditions. These breaches produce channels that will transport tides (figure 1.7). Under circumstances of a large *tidal prism* these channels can become permanent tidal inlets, which is how Hurricane Pass and Johns Pass were formed. Most of the time the tidal prism is too small to maintain the channel and longshore transport of sediment causes them to close.

Mixed-Energy Barrier Islands

All barrier islands are initially wave-dominated. It takes time and abundant sediment to eventually develop a mixed-energy barrier. As the name implies, this type of barrier island requires a combination of wave energy and *tidal flux* to form. These islands are quite common along Florida's Gulf coast. The reason is that some of the bays and estuaries landward of the barriers have a large

West Florida Inlets

Figure 1.8

Plot showing the tidal prism of inlets along the Gulf Coast of the Florida peninsula

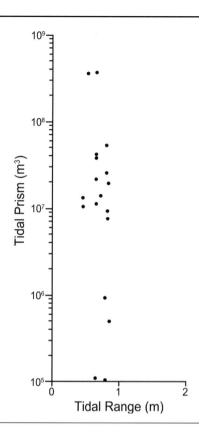

area. This produces a large tidal prism even though the tidal range is quite low. The volume of water in the tidal prism is calculated by the area of the bay served multiplied by the tidal range. On this coast that volume ranges over four orders of magnitude (figure 1.8). Regardless of the size of the tidal prism, the time available for the tide to pass through an inlet is dictated by the tidal cycle, which on this coast is *semi-diurnal* for most of the lunar cycle. That means the entire tidal prism must enter and leave through the inlet twice each lunar day. If the prism volume is large, it must move fast through a wide and deep inlet, and if the prism volume is small, it will be slow and move through a

narrow and shallow inlet. The speed of the tidal current controls the volume and rate of sediment movement and how the current interacts with longshore currents.

Tidal inlets can have large amounts of sediment associated with them on both the bay side and the Gulf side. The sediment on the open water Gulf side is called the *ebb shoal* (delta) and on the land side it is the *flood shoal* (delta) (figure 1.9). It can be large or small, with the influence of the tidal flux being the controlling factor along with sediment availability. It is typical that large tidal prisms cause development of large ebb shoals and that these shoals are commonly

upward/downward shoaling

shore-normal

Notes

Figure 1.9

Diagram showing sediment bodies associated with a tidal inlet

...bars of sand and their ...ement to the beach ...pturing sediment that ...ise be transported ...ier to distribute the ...venly. In this manner ... the *downdrift* end of the barrier to become sediment starved and the *updrift* end to prograde, forming what we call a *drumstick barrier island* (figure 1.11).

...pped at this end of the island. They merge with the beach and cause the island to prograde or increase in width.

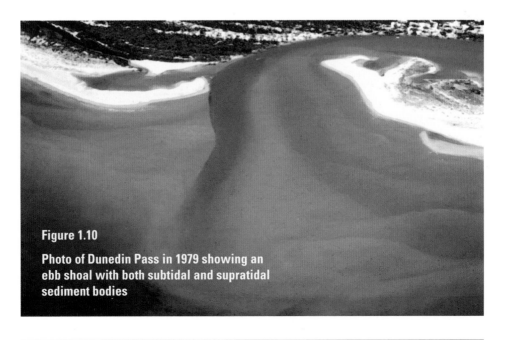

Figure 1.10

Photo of Dunedin Pass in 1979 showing an ebb shoal with both subtidal and supratidal sediment bodies

Barrier Island Drumstick Model

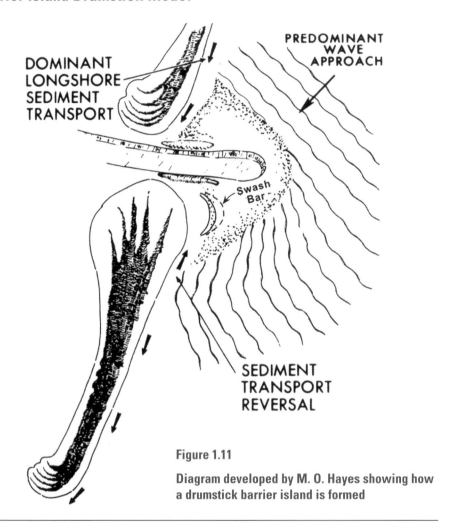

DOMINANT LONGSHORE SEDIMENT TRANSPORT

PREDOMINANT WAVE APPROACH

Swash Bar

SEDIMENT TRANSPORT REVERSAL

Figure 1.11

Diagram developed by M. O. Hayes showing how a drumstick barrier island is formed

The resulting mixed-energy barrier is common along this coast but not along other parts of the Gulf of Mexico coast. They are also common on the coast of the Carolinas and Georgia southward to the northernmost part of the east coast of Florida. The narrow end of a mixed-energy barrier is commonly washed over (figure 1.12) in a similar fashion to that of the wave-dominated barriers. The wide end of this type of barrier tends to have multiple beach-dune ridges as a result of the abundant accumulation of sediment (figure 1.13).

Age of Florida Peninsular Gulf Barriers

In order to understand the age of the barriers under discussion, it is important to understand how sea level has been behaving during the

Figure 1.12

Washover fans on the narrow end of Caladesi Island: a) an old and vegetated washover fan on the left, and b) a small, young sand fan at the arrow

Figure 1.13

Multiple beach-dune ridges on Little Gasparilla Island

past several thousand years. The melting of glaciers across the globe began about 18,000 years ago, producing a rapid rise in sea level. The ocean was about 350 feet below its present level when the glaciers were at their maximum. Sea level rose about a centimeter (2.54 cm = I inch) per year for several thousand years. This rate of change was so fast that barrier islands could not develop. About 7,000 years ago the rate of sea-level rise slowed dramatically, and the first barrier islands began to form in the Gulf of Mexico on the Texas coast. There were no barriers yet on the Florida coast.

Sea level reached almost its present position about 3,000 years ago, according to considerable research around the coasts of the United States. It was at this time that the first of the Florida Gulf coast barriers began to develop. The wide drumstick barriers such as Siesta Key date back to this time. The age of the barrier islands has been determined by radiometric carbon-14 dating of shell material and peat deposits in the oldest backbarrier sediments that are present on the islands. This process uses a radioactive isotope of carbon, ^{14}C, which can be used to date things as old as 30,000 years. This isotope is present in the shells and shell debris in the beach-dune ridges and in peat formed from mangrove and salt marsh accumulations, making this process easy. Specialty laboratories do this type of analysis routinely.

Dating the barrier islands on this coast to a maximum of about 3,000 years tells us that sea level has been at or near its present position for that length of time. There are actually three different theories on this situation. These are 1) that sea level has risen only a couple of meters over that time, 2) that sea level has moved slightly above and below its present position over that time, and 3) that sea level reached its present position about 3,000 years ago. The important condition is that sea level has been stable enough for the islands to form. This requires enough time to allow waves and wave-generated currents to accumulate a sufficient amount of sediment to develop the barriers. The fact that these Florida barriers are young compared to other parts

of the Gulf coast is due to the scarce supply of sediment on the adjacent *continental shelf*, which is dominated by limestone.

The Florida peninsula is a thick accumulation of limestone called the Florida Platform that dates back several tens of millions of years. For most of its history this platform was isolated and not connected to the mainland of the United States. It was only about 15 to 17 million years ago that the Florida Platform became attached to the continent. This permitted erosional debris such as sediment to be transported from the southeastern part of the continent southward to cover the Florida Platform. Little of this *terrigenous* material made its way offshore to the continental shelf. The environment of the Florida Platform was limestone rock with scattered shell debris, with little of the terrigenous sand that is necessary to form barrier islands.

Now that the rate of sea-level rise is considered to be increasing, coasts in general and barrier islands especially are in some jeopardy. The global rate of sea-level rise is about 3 mm per year, which is approximately the thickness of two pennies. In some parts of the Gulf coast the rate is three times that. Neither of those rates may seem alarming, but over a few generations it is a lot and can cause problems for coastal occupation. The rate of sea-level rise on the Florida peninsula will remain relatively low as compared to most other regions because it is based on limestone, which does not compact. Most trailing edge coasts on coastal plains are sinking due in part to compression of the thick sediment under them, which, when coupled with the global rate, produces the high rates of rise. It is likely that Florida's barrier islands will remain stable from sea-level change at least through the end of this century—if the rate of sea-level rise does not increase significantly.

Development of the Barrier Islands

For multiple reasons the Florida Gulf coast has become quite highly developed by human habitation. The appeal of the weather, the cost of living, fishing, and other water-oriented activities have taken the population of the coast from almost nothing at the turn of the twentieth century to several million at the present time. Barrier islands can maintain their nature and character when left in a natural state, but occupation by people can change that tremendously. Natural barrier islands can move landward or seaward, coping with waves, tides, storm impacts, and sediment supply. Some barrier islands may disappear when conditions become unfavorable, while new barrier islands may develop. Human occupation puts substantial limitation on the dynamic nature of barrier islands. Barrier islands are fragile systems that are home to an abundant and diverse community of plants and animals. The disturbance caused by human development changes so

much of the natural environment as to make it uninhabitable for native species. Construction of coastal protection structures, residential and commercial buildings, roads, causeways, and other infrastructure removes that natural condition of the barrier and prevents nature from taking its course.

The remainder of this book will look at this complex barrier-inlet system in detail to discuss the current conditions, ranging from natural to developed. The many barrier islands on the Florida Gulf coast will be grouped into categories of development or lack thereof. The islands and the tidal inlets that separate them will be discussed as systems because they range from completely natural to completely developed (figure 1.14).

Tidal Inlets

Barrier islands are the primary topic for discussion in this book, but the tidal inlets that separate them are quite important in the development and sustenance of the islands. They are an essential part of the barrier island system. It is important to have a general knowledge of these inlets in order to fully understand the barrier islands. In considering this barrier-inlet system as the most diverse in the world, it must also include the inlets.

Tidal inlets can be placed in three general categories: 1) tide-dominated, 2) mixed-energy, and 3) wave-dominated. The mixed-energy category is subdivided into

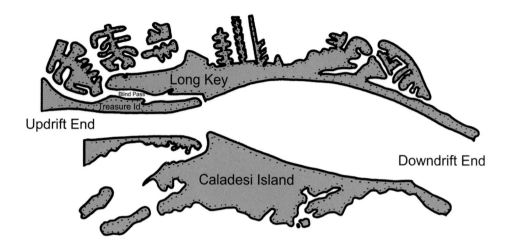

Updrift End

Downdrift End

Figure 1.14

**Outline maps of a natural barrier island (Caladesi Island)
and a heavily developed island (Long Key) on the Florida Gulf Coast**

two varieties: straight and offset (figure 1.15). The size and shape of the ebb shoal is what determines the inlet type. The most important factor that controls the size and shape of the ebb shoal is the tidal prism. It is the volume of water that flows in or out of a tidal inlet during a rising or falling phase of the tide. The primary factor that determines tidal prism on this coast is the size of the backbarrier water body, because tidal range is essentially the same. Tidal prism along this coast covers a range of four orders of magnitude (see figure 1.8).

Some inlets have closed during the past few decades because of reduction of the tidal prism. Engineers generally use a current speed of about one meter per second to keep an inlet open and stable. Aerial photos of some barrier islands show that there were tidal inlets during earlier stages of barrier development that have closed during prehistoric time. Two anthropogenic factors have caused the decrease in tidal prism that have resulted in the instability of tidal inlets. The most important is the extensive *dredge-and-fill* construction in the backbarrier bays (figure 1.16). Wetlands are dredged, with the spoil being placed adjacent to the dredged channels. This produces upland property that is used for development. Such activity causes reduction in the area of the bay and therefore the reduction in tidal prism. The other is when fill causeways are constructed to connect the barriers

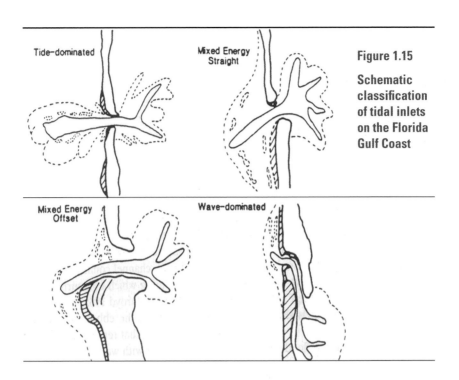

Figure 1.15

Schematic classification of tidal inlets on the Florida Gulf Coast

Tide-dominated

Mixed Energy Straight

Mixed Energy Offset

Wave-dominated

Figure 1.16

Aerial view of typical dredge-and-fill construction on Florida's Gulf coast

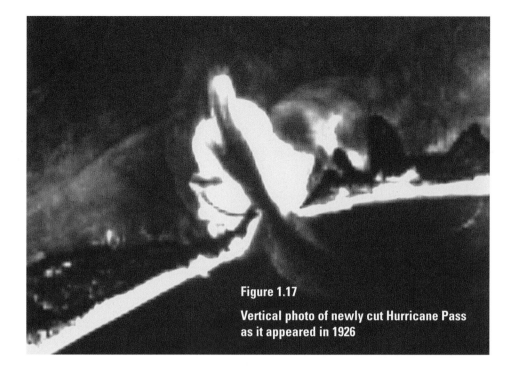

Figure 1.17

Vertical photo of newly cut Hurricane Pass as it appeared in 1926

to land or when a new inlet is cut by a hurricane (figure 1.17). In both of these cases the bays that service a particular inlet are reduced in size. It is quite apparent, therefore, that tidal inlets and barrier islands are an integrated coastal system. A natural factor that changes tidal prism is when a hurricane breaches a barrier island and forms a new tidal inlet. This means that the new inlet will capture part of the tidal prism from a previously existing inlet or inlets.

Chapter 2 How Barrier Islands Work

The previous chapter described barrier islands and provided a brief and simple classification. Basically there are only two types: wave-dominated and mixed-energy. There are no barriers on a tide-dominated coast because sand cannot accumulate in a shore-parallel fashion due to strong tidal flow. This chapter will first discuss the natural barrier island without any human impact. After that, the role that is and has been played by human activity on barriers islands

will be considered. All succeeding chapters will address both aspects of the barriers, beginning with the natural system and continuing with human influence on them.

The best place to begin this discussion is with a generalized map of a barrier island system (figure 2.1). This map includes all of the important elements of a barrier system, including the beach, dunes, washover fans, wetlands, and a tidal inlet with

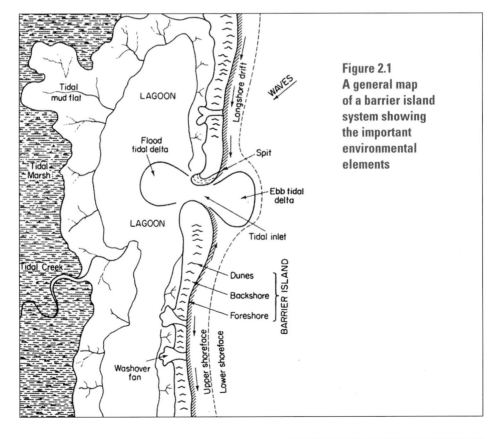

**Figure 2.1
A general map
of a barrier island
system showing
the important
environmental
elements**

both an ebb and a flood shoal. This gives some indication of how a complex group of environments is interacting to produce and modify a barrier-inlet system. The following discussion explains the process-response operations that form, maintain, and change the barriers and their adjacent inlets. This is collectively called morphodynamics.

The elements required to make this system are sediment—the material with which these depositional environments are constructed—and the processes to do the job, including wind, waves, and tidal flux. All of these vary within a wide range of speed, scale, and intensity. As was mentioned in the previous chapter, the initiation of barrier formation is typically through upward-shoaling of sand by wave action. As the linear, shore-parallel sediment body formed by waves is developed, it emerges to an intertidal and eventually a supratidal position relative to sea level (figure 2.2). Shortly thereafter opportunistic plants begin to colonize. This can take place in only a few months.

Once there are plants on this sand body and as the wind continues to transport sediment, sand begins to accumulate into coppice mounds, which are essentially incipient sand dunes (figure 2.3).

The other common process during this early stage of barrier development is the washing over of wave-generated flow that carries sediment and forms washover fans (figure 2.4).

As time passes the island grows and dunes become larger. Eventually the dunes are high enough in elevation so that only severe storms with large waves and elevated water level (storm surge) can transport water and sediment over the barrier. When that happens, the washover fans are larger because more energy is required to develop them on a

Figure 2.2
Sand bar showing the development of the beginnings of a barrier island through upward shoaling

Figure 2.3

Coppice mounds that are developing around vegetation on a bare sand surface

Figure 2.4

Thick washover deposits caused by Hurricane Andrew in 1992

relatively mature barrier. The other possible consequence of severe storms is the breaching of the barrier to form a channel through which tidal flux takes place (figure 2.5). These channels may persist, as in the case of Johns Pass and Hurricane Pass in Pinellas County. Johns Pass was formed by a hurricane in 1848 (figure 2.6) and Hurricane Pass was formed by a similar storm in 1921 (figure 2.7). In most of these situations, the channel formed by the storm closes within months or a few years due to longshore transport of sand, but both of those remain viable tidal inlets to this day. Willy's Cut, formed in 1984, closed in a little more than two years.

These are the process-response conditions that are formed directly by wind and waves. The other primary

Figure 2.5
The breaching of the narrow end of Caladesi Island in Pinellas County by Hurricane Elena in 1984 caused formation of Willy's Cut.

Figure 2.6

Photo of Johns Pass in 1945, a tidal inlet formed by a hurricane in 1848 that persists to the present

Figure 2.7

Photo of Hurricane Pass in 1926 after it was cut by a hurricane in 1921

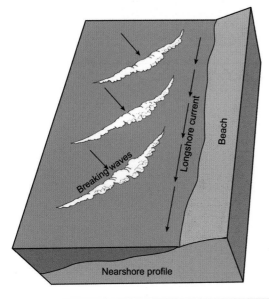

Breaking waves

Longshore current

Beach

Nearshore profile

Figure 2.8

Diagram showing the nature of shoreline currents generated by waves

coastal process that influences barrier-inlet morphology is the longshore transport of sediment (figure 2.8). This process is dependent on both the availability of sand and the wave climate at a particular time. The variables for waves include their size and their angle and direction of approach to the shoreline. During a storm it is common for the longshore current velocity to exceed one meter per second. That condition transports considerable sediment.

There is one more coastal process that is common along barrier island coasts but that is not important in barrier island morphodynamics. That is rip currents. These are narrow currents that move seaward from near the shoreline through the surf zone where waves break. They occur most frequently during conditions of waves approaching the coast parallel to the shoreline (figure 2.9). Waves do transport some water shoreward. It tends to "pile up" at the shoreline but must return seaward to try to stabilize the water level. It does so in narrow currents that are commonly a problem for swimmers, but these currents typically do not transport significant amounts of sand.

Barrier Island Morphodynamics

Now that we have explained how individual environments develop and can be changed, we can move on to how the system operates as a whole. Each of the two primary barrier types will be considered separately. The principal morphodynamics are the same regardless of the age or size of the barrier and the adjacent inlet. The reader should know that like all classifications there is a continuum of change from one category to another. Sometimes the category into which an individual barrier is placed is a difficult choice. This will be apparent in the discussions of the individual barriers in the following chapters. Regardless of the age or size of the barriers, the wave and tidal processes are the same.

Figure 2.9

Diagram showing how rip currents are generated by onshore transport of water by waves

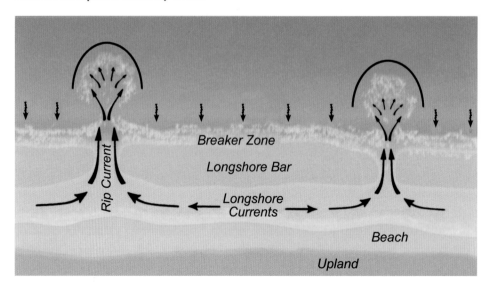

Wave-Dominated Barriers

These barriers are narrow and long with small, typically unstable tidal inlets at one or both ends (figure 2.10). Washover fans tend to be present throughout the barrier because the elevation is low and the island is narrow. The occurrence of these washovers is limited by the elevation. In general the young and therefore low elevation barriers are washed over throughout (figure 2.11). The more mature ones tend to be infrequently and locally washed over or even not at all if their dunes are large (figure 2.12).

Wave-dominated barriers depend on a large sediment supply and wide beaches to maintain their size and to "grow." The wide beach provides a supply of sand for development of dunes, and as the dunes become higher, washover is decreased. If dunes stay low and washover takes place on a regular basis, the barrier will not increase in width or elevation; it will migrate landward instead.

The ends of a wave-dominated barrier may experience three different conditions as time passes: 1) they may continue to extend the length of the island (figure 2.13); 2) the strong tidal currents may transport the sand being provided by longshore currents to the end of the island and across the small and unstable tidal inlet (figure 2.14); or 3) the end may make essentially a right-angle turn

Figure 2.10

View of Manasota Key, a long and narrow, wave-dominated barrier island

Figure 2.11

Photo of a low elevation barrier spit showing abundant washover deposits

Figure 2.12
Mature, relatively high elevation wave-dominated barrier that is rarely washed over or breached

Figure 2.13

Barrier that is being extended by longshore transport of sand to the right (south)

into the tidal inlet (figure 2.15). Each of these scenarios is dependent on sediment supply, longshore sediment transport, and tidal currents.

Mixed-Energy Barriers

Developing a mixed-energy morphology requires a more complex morphodynamic system than is the case with wave-dominated barriers. The tidal inlets play a more important role in the development of the overall shape of the barrier. A key to the morphology is the presence of a well-developed ebb shoal. This ebb shoal (figure 2.16) causes waves to refract around it and produce a local reversal in direction of the longshore transport of sand, as shown in figure 1.10 (page 15). This local reversal essentially traps sediment in the updrift portion of the barrier, keeping it from continuing in the longshore transport system. The result is that this trapped sediment accumulates

in swash bars (figure 2.17) that eventually migrate landward and become beach-dune ridges (figure 2.18). This is how the wide end of the mixed-energy barrier develops and is why the name "drumstick barrier" has been applied. This end of the barrier may include many of these beach-dune ridges. In some cases the low, linear area between adjacent ridges may have water in it. These narrow water bodies are called catseye ponds (figure 2.19) because of their long and linear shape.

The other end of the mixed-energy barrier is essentially sediment-starved because of the trapping of the sediment on the updrift portion. Once the ebb shoal develops, the only sediment that reaches this part of the island is from offshore or the middle of the island shoreline area. The result is that this end of the island is narrow, has a low elevation without

Figure 2.14

One wave-dominated barrier that has been extended to the right (south) by longshore sand transport, closing the inlet that was unstable

Figure 2.15

Wave-dominated barrier that has a tidal inlet interrupted by the longshore transport of sand causing it to make a right-angle turn to parallel the inlet channel

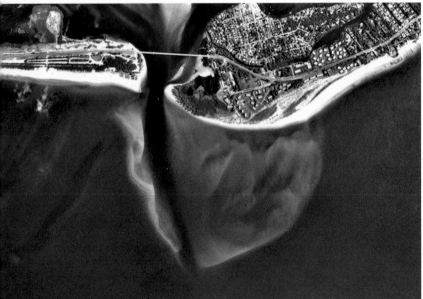

Figure 2.16

Aerial photo of a well-developed ebb shoal that would cause wave refraction and a local reversal of sediment transport

dunes, and is washed over by even modest storms. It is characterized by an eroding open water side (figure 2.20a) and washover fans on the backbarrier side (figure 2.20b).

Because of the dearth of sand and the low elevation of this end of the drumstick barrier, erosion is common and typically results in deposition of washover fans. Multiple occurrences of this process cause the barrier to migrate landward or to be destroyed. Remember that the backbarrier portion of these barriers

Figure 2.17

Intertidal swash bar that develops by wave action adjacent to a tidal inlet and will migrate to weld onto the beach within some weeks.

Figure 2.18
Oblique view along a mixed-energy barrier island showing multiple beach-dune ridges in the foreground

is characterized by mangroves, especially the red mangrove (*Rhizophora mangle*). These are very sturdy trees with large and deep roots. They drop leaves regularly, forming a thick accumulation that eventually becomes mangrove peat. When the Gulf side erodes, it exposes this mangrove peat as shown in figure 2.21. It can become buried if a sand supply becomes available.

The Big Picture of the Florida Gulf Peninsula Coast

Some important trends and generalizations can be made about this complex barrier-inlet system. The tidal range is pretty uniform

Figure 2.19

An example of the long and narrow water body referred to as a catseye pond

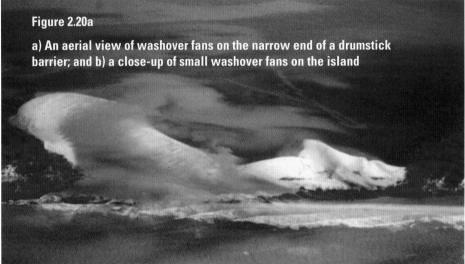

Figure 2.20a

a) An aerial view of washover fans on the narrow end of a drumstick barrier; and b) a close-up of small washover fans on the island

Figure 2.20b

Figure 2.21

Exposed mangrove peat and relict mangrove trunks on the eroding open-water side of the narrow end of a drumstick barrier

along the entire coastal reach, with a spring range of less than 3 feet. It actually increases both to the north and south beyond where barrier islands are present (figure 2.22). Wave energy is also both low and uniform—a mean annual significant wave height of about 1.0–1.5 feet.

Regional Longshore Sediment Transport

The prevailing direction of approach of the waves is from the southeast, which causes a south to north direction of longshore sediment transport. The overall predominant direction of longshore sediment transport is from north to south. Recall that "prevailing" refers to time and "predominant" refers to energy. There are several places where the predominant direction of sediment transport is from south to north. This is generally due to the orientation of the shoreline of a particular island or portion of an island.

The longest stretch of the coast where longshore sediment transport is south to north includes much of Clearwater Beach Island and all of Caladesi Island into Hurricane Pass (figure 2.23). Another obvious example of direction reversal is on Sand Key, where the shoreline trends to the mainland on both ends with longshore transport diverging at the center of the island (figure 2.24, page 38). Other places where the net annual sediment transport is south to north include portions of Honeymoon Island, Casey Key, and Estero Island. Figure 2.23 shows the direction of sediment transport along the entire northern portion of the barrier island system from Anclote Key to Venice, where the barriers do not exist for a few miles.

The rate of longshore sand transport can be high in terms of the gross amount, but the net rate is low. We can make a comparison with the east

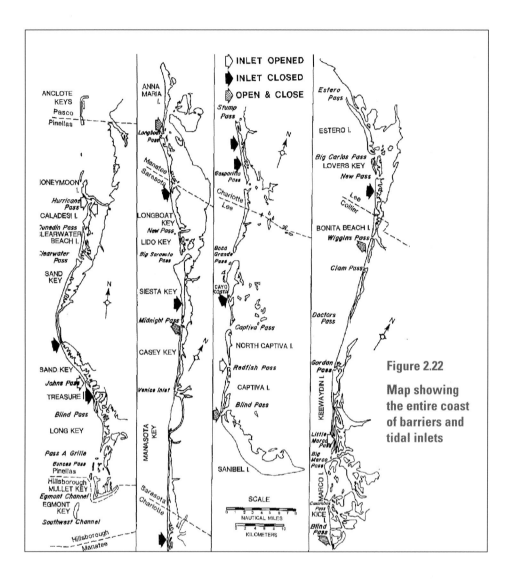

INLET OPENED
INLET CLOSED
OPEN & CLOSE

Figure 2.22

Map showing the entire coast of barriers and tidal inlets

coast of Florida, where the annual rate of longshore sediment transport starts at about 500,000 cubic yards per year in the north and decreases to the south. On the Gulf side of the peninsula, the net annual rate is less than 100,000 cubic yards everywhere, and not even half that amount on most of the coast. The gross rate is much higher because of the distinctly bimodal transport. The prevailing southeasterlies cover a long time and transport at a high rate, and the predominant wind with the frontal passage during the winter also transports at a high rate, but over a short time. The difference is only tens of thousands of cubic yards, mostly toward the south.

Location of Gulf Peninsular Barriers

The locations of the barrier islands on this coast are somewhat unusual. It was mentioned earlier that the

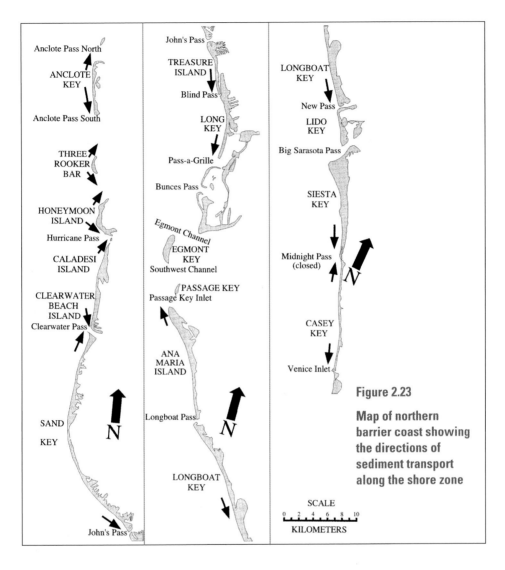

Figure 2.23

Map of northern barrier coast showing the directions of sediment transport along the shore zone

SCALE
0 2 4 6 8 10
KILOMETERS

Florida peninsula is a carbonate platform that is about a mile thick dating back more than 75 million years. The youngest of these carbonate strata are only a few million years old. These are Miocene age strata that have more recently been covered with quartz-rich sediment that originated in rocks of the Appalachian Mountains and was eroded and transported to the coast. It was then carried south by waves and currents to eventually cover the Florida peninsula.

The Miocene strata are carbonate in composition and are exposed at the surface or only slightly below the surface. They can be seen exposed along the coast in question near Tarpon Springs, Indian Rocks, and at Venice. Research using cores of sediments and sedimentary rocks shows that these Miocene strata are

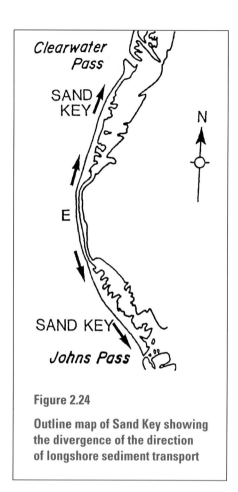

Figure 2.24

Outline map of Sand Key showing the divergence of the direction of longshore sediment transport

located only a few feet below sea level at some locations. In such a position this bedrock provides an antecedent topography that has influenced the position of Florida Gulf barrier islands along this coast. One of the best places to see this condition is at the northern portion of the barrier system in Pinellas County, from Anclote Key to Sand Key (figure 2.25). Anclote Key is more than a mile from the mainland. Moving to the south the island becomes closer until the Indian Rocks headland at Sand Key.

Many sediment cores from this area permit construction of a sequence of diagrams to show how these barriers developed and how the underlying Miocene limestone provides the antecedent topography to "anchor" the barrier islands where they are located (figure 2.26). As sea level was rising it is probable that barriers

Figure 2.25

Aerial photo of the north portion of the barrier-inlet system showing its position relative to the mainland shoreline

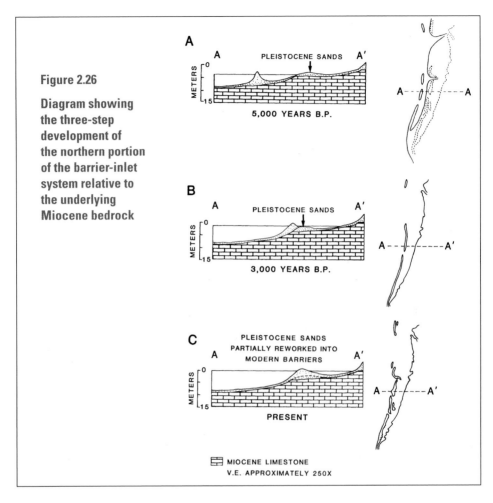

Figure 2.26

Diagram showing the three-step development of the northern portion of the barrier-inlet system relative to the underlying Miocene bedrock

began to form about 5,000 years ago and were then destroyed as sea level continued to rise until the change in slope of the limestone caused the sand body to accumulate and grow at the present location. A close-up diagram shows this relationship at Hurricane Pass with Honeymoon and Caladesi islands on either side (figure 2.27).

Moving further to the south to the headland area at Indian Rocks the barriers approach the mainland. Here the total distance from the rocky headland across the bay and

the barrier island is only about 200 yards (figure 2.28).

A somewhat different situation occurs at Venice near the middle of the barrier system (figure 2.29) just south of the Venice Inlet. Here the bedrock mainland reaches the shoreline. No bay or barrier island is present. A canal was dredged to allow the Intracoastal Waterway to continue (figure 2.29). At Manasota Key the typical barrier and bay situation begins again. The barrier island along this reach of coast south of the Venice headland is typical,

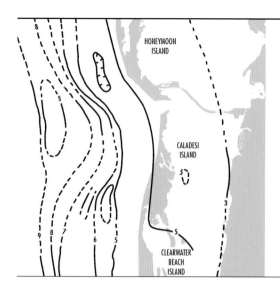

Figure 2.27

Close-up map of the shoreline of Honeymoon (top) and Caladesi (bottom) islands relative to the antecedent topography of the underlying Miocene bedrock

Figure 2.28

a) Aerial photo of the Indian Rocks headland area and
b) a close-up of the mainland coast that is Miocene bedrock

Figure 2.29

Aerial photo of the Venice area showing the absence of a barrier south of the Venice Inlet and a continuation again at Manasota Key

with a well developed bay system and barriers separated by tidal inlets.

The next significant change in the nature of the barrier system is at the south end of Sanibel Island where there is a major change in the shoreline position (figure 2.30).

The largest change in shoreline is on Sanibel Island, where there is a major shoreline dislocation (figure 2.31). This is probably due to the underlying trend in the pre-Holocene substrate. Much like the areas to the north, the antecedent topography caused by Miocene limestone has caused the

barrier system from Estero Island to the south to be located where it is. Tectonic activity in the Charlotte Harbor area has caused dislocation of the antecedent topography base on the Miocene limestone. That is the reason that the barrier island system has been dislocated as shown in figure 2.31.

Summary

The morphodynamics of barrier islands is generally simple, but it can be complicated as the variables change through both space and time. The processes and the sediment that

Figure 2.30

Sanibel Island exhibits a very large change in the shoreline orientation

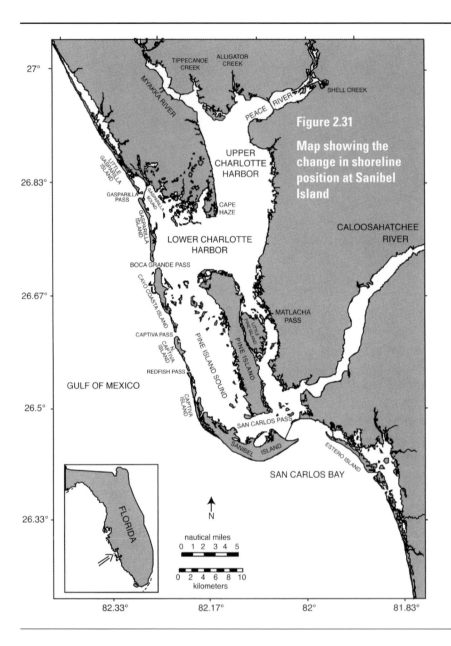

Figure 2.31

Map showing the change in shoreline position at Sanibel Island

they act upon are considered to be in the generally low-energy level due to the small tidal range and low mean annual wave height along this coast. In spite of that, changes can take place rather rapidly. Storms can make major changes in only hours or days. Inlets carry a wide range of tidal prisms that are major factors in influencing the morphodynamics of the system. The large scale of sediment transport shows most of this coast has a north to south direction, but there are many local places where it is the reverse.

Chapter 3 Pristine Wave-Dominated Barriers

There are two distinct types of wave-dominated barrier islands that are essentially undisturbed. The basic difference is size and age. They group nicely because the short ones are also the young ones. They are similar in their morphology as well as their pristine character. Because of their important differences, each of these types will be discussed separately. In general, barriers in this overall category are public land, unoccupied or scarcely populated, inaccessible by vehicle, and very popular for daily recreation, including overnight camping.

Small Young Barriers

This group of barrier islands is concentrated on the north end of the Florida Gulf peninsula, although they also occur in the southern portion. The youngest islands are all only a few decades old and about a mile or two long. The prominent young ones are Three-Rooker Bar, Shell Key (previously called North Bunces Key), and South Bunces Key, all in Pinellas County. At the present time South Bunces Key has been destroyed. Two small barriers in other locations are in their early stages of development. We'll call them North Anclote Bar and South Anclote Bar. Overall this part of the coast is sediment starved in the inner continental shelf. Of these barriers Three-Rooker Bar is near the north end of the island system, whereas Shell Key is near the huge ebb shoal at the mouth of Tampa Bay. The lack of sediment is present on the north portion of this region, but near the mouth of Tampa Bay there is abundant sediment.

Three-Rooker Bar

Three-Rooker Bar (figure 3.1) started to express itself as an intertidal

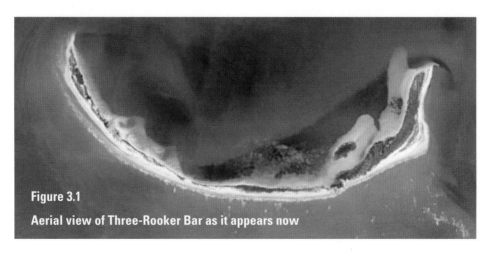

Figure 3.1

Aerial view of Three-Rooker Bar as it appears now

shoal in the late 1960s and its first supratidal expression was in 1973 (figure 3.2a). As time passed into the 1990s, it continued to grow (figure 3.2b). Available sediment was quite limited through the early development of this part of the coast. The reason for this condition was the presence of an extensive seagrass community in the inner shelf just beyond the narrow surf zone. In the early 1960s something, as yet unresolved, caused the seagrass to die. As a result, the sediment present and being stabilized by the grass was mobilized by waves and wave-generated currents. It was transported to the shallow water where it accumulated. An intertidal sand bar and eventually a small island formed—what is now known as Three-Rooker Bar. Other elements of this barrier system were also influenced

by this release of the stabilized sand and will be discussed later.

As time passed in the late twentieth century, Three-Rooker continued to grow from sand being added from offshore (figure 3.3). Severe storms caused washover fans, transporting sand over the island and into the backbarrier area. Erosion took place in the central part of the island where wave energy was highest (figure 3.4).

The area near the main channel into Tampa Bay is bounded by a huge sand body, one of the largest that extends into the Gulf of Mexico (figure 3.5). This sand body provides sediment for the young barriers that develop in this area.

Shell Key (North Bunces Key)

North and South Bunces Key near the mouth of Tampa Bay were developed

---

Figure 3.2a

Initial development sequence of Three-Rooker Bar taken from a sequence of aerial photos

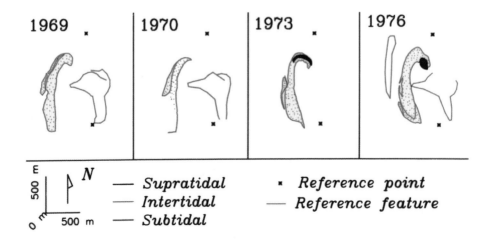

Figure 3.2b

Subsequent development sequence of Three-Rooker Bar

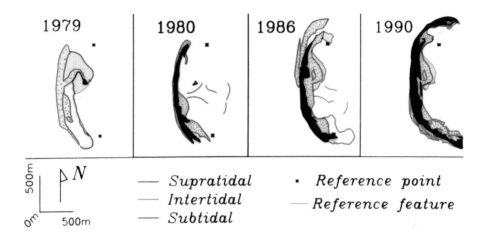

| 1979 | 1980 | 1986 | 1990 |

N
500m
0m   500m

— *Supratidal*   • *Reference point*
— *Intertidal*   — *Reference feature*
— *Subtidal*

Figure 3.3

South end of Three-Rooker Bar showing washover sediment landward of the barrier and a swash bar welding on to the barrier on the Gulf side

by processes similar to those that formed Three-Rooker Bar, but under circumstances of abundant sediment. All were developed from sediment associated with the ebb tidal shoal at Bunces Pass (figure 3.6). Shell Key (North Bunces Key) (figure 3.7) formed in the early 1960s, about fifteen years before South Bunces Key. Swash bars formed and migrated to the main sand body. Storms breached Shell Key (figure 3.8), but it has persisted for more than fifty years. It developed and was

Figure 3.4

An erosion scarp near the center of the island where waves tend to be concentrated

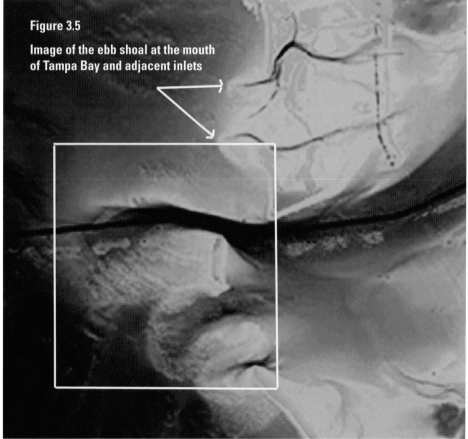

Figure 3.5

Image of the ebb shoal at the mouth of Tampa Bay and adjacent inlets

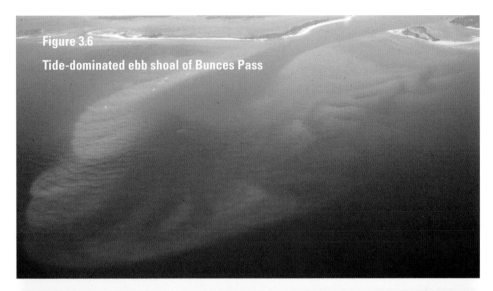

Figure 3.6

Tide-dominated ebb shoal of Bunces Pass

Figure 3.7

Early stage of the development of Shell Key (North Bunces Key) that has already been breached by a storm

breached, then the breach was sealed (figures 3.9).

South Bunces Key

South Bunces began in the same fashion as its neighbor (figure 3.10). Over its short lifetime it has experienced considerable growth (figures 3.11 and 3.12). As time passed, South Bunces Key migrated landward (figure 3.13). This, combined with erosion, progressed

Figure 3.8

More mature Shell Key with extensive vegetation. The breach persists at this time (about 1990).

Figure 3.9

Shell Key as it currently appears with the breach closed

to the point that this island became unrecognizable (figure 3.14). Now it has been completely destroyed and the main part of Mullet Key is being eroded (figure 3.15).

These small and low-elevation wave-dominated barriers are vulnerable. Only a meter or two of storm surge will overtop them and cause washover. The waves that are superimposed on top of this surge can also cause considerable erosion. This means that a moderate to severe hurricane could completely destroy these islands.

Figure 3.10

Emerging sand shoals in 1977 that were the beginning of South Bunces Key that fronted Mullet Key

Figure 3.11

Developing South Bunces Key several years later as vegetation began to establish itself

Older Pristine Wave-Dominated Barriers

Anclote Key

Probably the most common type of barrier island along this coast is the wave-dominated barrier. Some, such as Anclote Key (figure 3.16), are pristine, while others, such as Sand Key (discussed in a later chapter), are completely developed. Such barriers are present throughout this barrier island system. The most studied of the pristine barriers is Anclote Key, the northernmost of the barrier islands in this extensive chain. It is about 3 miles (5 km) long with shallow and wide tidal channels at each end. In addition, there are two

Figure 3.12

Looking Gulfward across South Bunces Key, which has attached to Mullet Key at both ends as it appeared in 1990

Figure 3.13

South Bunces Key has begun to migrate landward toward Mullet Key as storms cause erosion and washover events.

Fig. 3.14

Oblique aerial photo of the north portion of Mullet Key where the small barrier of South Bunces Key was prior to this time

Figure 3.15

The current situation on Mullet Key, where all of South Bunces Key has been eroded and transported to the south and Mullet Key itself is eroding as it appears in this early 2015 photo

Figure 3.16

North end of Anclote Key as it appeared in 1979

incipient barriers developing at each end, North Anclote Bar (figure 3.17) and South Anclote Bar (figure 3.18), respectively.

Based on carbon-14 dating of material from sediment cores, Anclote Key is about 1,200–1,500 years old. As far as is known from many of these sediment cores, it has

always been wave dominated. Some of these sediment cores penetrate to the Miocene limestone that provides the antecedent topography that positions the barrier where it is, far from the mainland. The present morphology includes extensive mangrove stands on the landward side of the island. This shoreline is irregular in outline, indicating that

Figure 3.17

Anclote Key with a small
emerging shoal off its north end

Figure 3.18

South end of Anclote Key with considerable sand in
the form of swash bars merging with the island

Figure 3.19

Infrared aerial photo of Anclote Key as it appeared in 1973

it is the result of washover sediment deposited by storm surge and waves that overtopped the low island (figure 3.19).

Washover events on this island have been infrequent over the past several decades except for the northern portion of the island. As the island matured and as the beach accreted and became typically wide, considerable sand was transported landward by onshore wind. This resulted in the development of dunes up to 10 feet high. These dunes, along with the lack of hurricanes and storm surge, have prevented washover events.

Major changes have taken place along this portion of the barrier coast over the past five decades. Aerial photos from the 1950s show seagrass beds in shallow water just seaward of the surf zone along the northern Pinellas County coast. This seagrass community lasted until about 1960 based on aerial photo collections at the Florida Department of Transportation. An extensive study of what happened to the seagrass was undertaken by the author as part of an earlier study. Possibilities included harvesting by sea urchins, water temperature increases due to the nearby power station, water quality issues, and removal by storm waves. The sea urchin population was not great and biologists do not believe that they could destroy the entire seagrass community. The water temperature data from the effluent at the power station are available and do not show significant temperature increases. There is nothing to indicate that water quality was diminished. Hurricane Donna was the only significant storm of that time (1960). Although it was a major storm, it traveled east of the Gulf coast over land as it passed the Anclote Key area with strong offshore wind. It is very unlikely that such wind would develop waves in the Anclote area large enough to uproot and destroy the seagrasses. The bottom line is that we do not know what caused the demise of this sea grass community.

Regardless of the cause of the loss of the seagrass, it has had a major effect on Anclote Key and the adjacent coast, and the effect continues to today. The Anclote Key of the 1950s and 1960s had a recurved south end and a north end that terminated in shallow sediments (figure 3.20). Figure 3.20 shows both of the small and shallow tidal channels at each end of the island. They will be

discussed later in this chapter. The general effect of the absence of the seagrass is to cause the nearshore and inner shelf sediment to become mobilized, even by the low wave energy of this coast. This portion of Pinellas County experienced a large volume of landward sediment transport. It influenced multiple barrier islands.

The general direction of sediment transport along this part of the coast is south to north. As the sediment was transported to the beach of Anclote, it was also moved northward along the barrier. This resulted in a series

Figure 3.20

Outline maps of Anclote Key showing a century of change beginning with the map of 1883

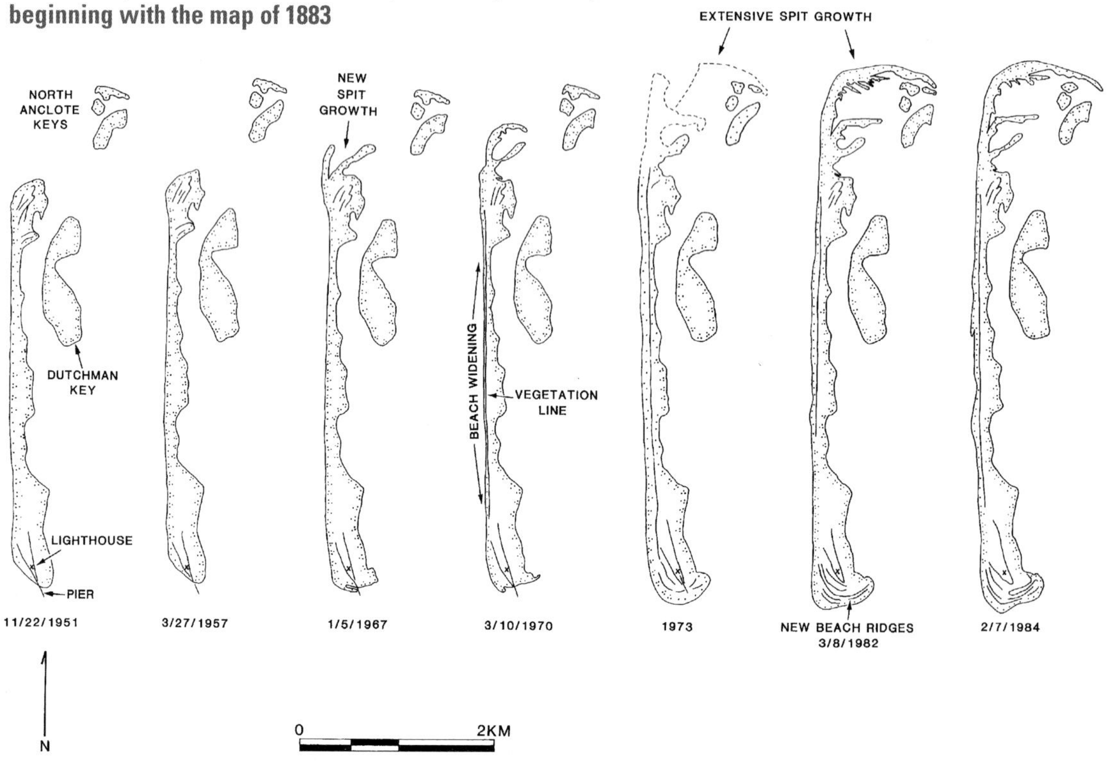

EXTENSIVE SPIT GROWTH

NORTH ANCLOTE KEYS

NEW SPIT GROWTH

BEACH WIDENING

VEGETATION LINE

DUTCHMAN KEY

LIGHTHOUSE

PIER

11/22/1951   3/27/1957   1/5/1967   3/10/1970   1973   NEW BEACH RIDGES 3/8/1982   2/7/1984

N

0   2KM

Figure 3.21a

a) Photo showing an anchored sailboat (arrow) landward of the north end of Anclote Key in 1973 and b) the same area as it appeared in 1979 showing the accretion at the north end of the island

Figure 3.21b

of *recurved spits* on the north end of Anclote Key (figure 3.21). It is important to grasp how much this end of the island changed in less than a decade with very low wave energy and slow longshore currents. Prior to the addition of the new sediment it was possible for a medium-sized sailboat to enter the quiet water behind the north end of Anclote for safe anchorage (figure 3.21a). Only a few years later this entry was blocked by the deposition of new spits at the north end of the island positioned by the presence of the tidal channel (figure 3.21b).

This new north end of Anclote was initially unvegetated, but in only a few years opportunistic plants and small trees were sprouting up along its entire length. These low

Figure 3.22

Close-up of the stratification in washover sediments after Hurricane Elena in 1985. The dark layers at the bottom are algal mats over which the sediment was deposited.

spits were vulnerable to washover activity during storms. Such actions widened the spits and carried sand into the backbarrier waters. The washover sediments displayed good stratification, which would last only until the roots of vegetation destroyed it (figure 3.22).

In the late 1990s another significant influx of sediment appeared in the Anclote Key area. It came in two places: on the southern portion of the barrier and just off the northern end. A large intertidal sand bar moved into the intertidal position on the south, and as sediment was added it became supratidal (figure 3.23). This bar has been enlarged and extended so that it has closed off the small tidal channel. It is now vegetated and has been termed South Anclote Bar (figure 3.24). On the other end of the barrier another sand bar has developed. This sand body developed and was removed

multiple times in the 1990s and has now enlarged and become vegetated to the point that it has also been named: North Anclote Bar.

The Anclote Key area has become one of the most dynamic of the entire Florida Gulf peninsular barrier system. The island has been enlarged and new sand bodies have become substantial in size and have cut off both of the tidal channels at either end of the barrier.

Lovers Key

A small, pristine, wave-dominated barrier island is immediately south of Estero Island and Ft. Myers Beach in Lee County. It is Lovers Key and it contains a state park with typical amenities as its only development. The island is only about 1.5 miles long and has right angle changes in orientation at each end (figure 3.25). The island is dominated by wetlands that extend well back into the *backbarrier estuary*. The Gulf side

Figure 3.23a
Aerial photo showing large sand bar just offshore of the south end of Anclote Key

Figure 3.23b
Ground view of ridge and runnel sand body that is slowly making its way to the south end of Anclote

of the island includes a sand beach (figures 3.26a and 3.26b), and dunes are essentially absent.

Because of its low elevation and lack of significant protection from a wide beach, this barrier is quite vulnerable to tropical storms and hurricanes.

Keewaydin Island

This island is unusual in its morph-ology and demographics. It is seven miles long and is located between Naples and Marco Island (figure 3.27). Keewaydin is being included here as pristine although there are some fifteen residences on the island. There is no commercial enterprise and there are no roads or vehicular access. Much of the island is a nature preserve. The homes are generally

Figure 3.24

Oblique aerial photo showing a sand body that is emerging off the north end of Anclote Key, extending the island to the south

Figure 3.25

Oblique aerial photo of Lovers Key with extensive wetlands on the landward side and New Pass in the foreground

Figure 3.26a

Accretional beach on Lovers Key at the state park

Figure 3.26b

Erosional beach on Lovers Key at the state park

large (figure 3.28) and owned by wealthy people who do not live there but visit occasionally.

Keewaydin Island is long, narrow, and distinctly wave-dominated. Unlike other similar morphologies along this coast, it does not display any coastal reaches that are erosional. On the contrary, the beaches of this island are stable and relatively wide (figure 3.29). A look at the southern portion of the island shows distinct

Figure 3.27

Looking to the south along the coast of Keewaydin Island toward Naples

Figure 3.28

View across Keewaydin Island showing one of the large homes on the island

beach ridges that accreted to the island and extended it (figure 3.30). They demonstrate that the net sediment transport on the island is to the south, as it is along most of this coast. The absence of erosion on this totally natural shoreline brings up the question of the negative influence of any hard structures on shorelines. Keewaydin Island is a great example of how well a natural shoreline can perform along a coastal region that is otherwise replete with structures and erosion.

Kice Island and Morgan Island

The southernmost barrier in this system is also a pristine, wave-dominated one. It is unusual in that it can just barely be called a barrier island. It is a complex environment

Figure 3.29

View of the excellent beach on Keewaydin Island

Figure 3.30

Looking south along Keewaydin Island showing accreting beach ridges that demonstrate a dominant sediment transport to the south

of a narrow beach with extensive mangrove mangals (figure 3.31). Although originally a single coastal element, it is now separated by Morgan Pass, and now each section of the island carries its own name. There are and have been a few homes on the property, but they have either been removed by shoreline retreat or they are in jeopardy.

The Morgan (south) portion of this element of the coast shows considerable loss of beach and sand over the past decade, especially on its southern end. Tropical storms and hurricanes from 2005 up to 2013 have removed much of the beach and caused landward migration of the southern half-mile or so toward the shore. Hurricane Wilma passed across the Collier County coast on October 21, 2005, and had a major impact on some beaches but not on others. This storm has been well documented by the Florida Bureau

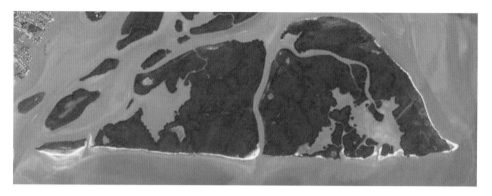

Figure 3.31

Vertical image of Kice-Morgan Island from Caxambas Pass on left to the tip of the barrier on the right as it existed in 2014

of Beaches and Coastal Systems. It had little negative impact on Keewaydin to the north and only modest impact on Kice Island and Morgan Island (figure 3.32). In these pre- and post-Wilma photos, the only change along the open coast is some landward transport of beach sand, but not a major change in shoreline position. Considerable sand has moved southward along Morgan Island, forming an accreting spit (figure 3.33).

The Cape Romano portion of Morgan Island has experienced a major change in the position of its shoreline. A look at this spit shows the beach ridges that developed this portion of the island. Another series of storms subsequent to Hurricane Wilma has caused major erosion of this spit area through a combination of washover and longshore transport of sand. The effect on an odd housing complex here has been well documented as these coastal changes have taken place. This group of adjacent concrete dome structures was constructed for use as a home in 1981. As time passed and the shoreline changed, it was abandoned in 2007 and fell completely within the nearshore environment by 2013 (figures 3.34 and 3.35). Much of this sediment has passed the end of Cape Romano and is now incorporated in the tide-dominated, shallow shoals that occupy the open water environment in this area.

Summary

Unfortunately for the coast, the few described in this chapter are the only pristine wave-dominated barrier islands along the Gulf coast of the Florida peninsula. Their location away from the mainland and, for the most part, inaccessible by vehicle helps to preserve their character. The only one that can be reached easily is South Bunces Key at Fort De Soto Park, and it has lost its character due to erosion and landward migration. The shorelines of these barriers range from accretional (growing) through

September 29, 1999

October 25, 2005

Figure 3.32 (left)

View of beach area on Kice-Morgan Island before and after Hurricane Wilma in 2005

Figure 3.33 (below)

View to the Gulf across Cape Romano showing abundant sediment on the southern end and a discontinuous beach to the north as it appeared in early 2000s. The dome buildings are at the long arrow and another home is indicated by the short arrow.

Figure 3.34

Oblique look at Cape Romano showing the position of the dome buildings (long arrow) and the home that is now well into open water (short arrow) as they appeared in 2005

Figure 3.35

Position of the dome buildings in 2013 when all of them were in open water as the result of erosion of the shoreline between 2005 and 2013

stable to erosional. As would be expected, the major changes are related to intense storms, but the nature of the changes themselves varies among the islands. It is also important to recognize that major changes of an accretional nature can take place in decades, a figurative nanosecond in geologic time, with Anclote Key being a prime example.

The preservation of these pristine barrier islands is important for all of us to see how nature works to form them. As you read further chapters that describe the extent and nature of human development on these islands, you will see how that activity conflicts with the pristine nature of the system.

Chapter 4 — Moderately Developed Wave-Dominated Barriers

There's a fine line between the various categories of development on the individual elements on this barrier coast. Multiple aspects of development must be considered. Are park-related buildings and other structures considered, and if so, to what level? What about infrastructure? Is that also included? Houses and commercial buildings are no-brainers. It's challenging to define what we mean by "moderately developed" barrier islands. In at least one case included here, an island was developed in the past but is now almost pristine. For our purposes, this category includes barrier islands with mostly residential development and little or no commercial development.

Mullet Key

Adjacent to the north shore of Egmont Channel, the entrance to Tampa Bay, is a wave-dominated barrier called Mullet Key. This is an unusual barrier in that it has a right-angle morphology (figure 4.1). This shape is due to the combination of wave processes on the west or Gulf side and tidal processes on the Egmont

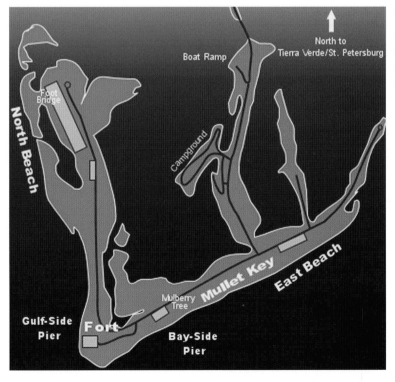

Figure 4.1a

Schematic map of Mullet Key showing main elements

Figure 4.1b

Oblique aerial photo of the Fort De Soto area

Channel side. This shape is unusual on the Gulf peninsular barrier coast. It is the result of longshore currents and onshore waves moving sediment shoreward and alongshore on the Gulf side. This sediment is captured by flooding tidal currents into Tampa Bay that carry it along the channel margin.

Mullet Key is a county park without any residential or significant commercial enterprises. It has extensive roads, an RV park, large parking lots, snack bars, change buildings, and a historic fort, Fort De Soto (figure 4.2). It is for these reasons that the island is in

this moderately developed category. Virtually nothing in the various types of development has an impact on erosion or any other morphologic changes.

The most important morphologic change that has taken place on Mullet Key in several decades is the development of what has been called South Bunces Key just offshore of the Gulf side of the barrier (figure 4.3). This sediment body emerged to intertidal level as the result of wave action in 1977. As time passed sediment continued to accumulate and a small barrier formed. In a

Figure 4.2a

Gun emplacement at Fort De Soto

Figure 4.2b

Buildings at Fort De Soto

way, it mimicked the barrier north of Bunces Pass, North Bunces Key (Shell Key), discussed in the previous chapter. Eventually South Bunces Key became vegetated and was the recreational beach of choice for Mullet Key (figure 4.4). After ten years, the barrier migrated somewhat toward Mullet Key proper (figure 4.5).

This condition continued for a few decades, but more recently conditions on this barrier have changed markedly. Storm activity over the past decade has caused

Figure 4.3

Emerging sand bar that eventually becomes South Bunces Key as it appeared in 1977

Figure 4.4

South Bunces Key as it appeared about 1980 with a lagoon landward of it

South Bunces Key to erode and move landward to merge with Mullet Key so that the former small barrier is no longer recognizable (figure 4.6). In addition, there is severe erosion along this portion of Mullet Key (figure 4.7). Only a short distance to the south on this part of the barrier, however, the beach is very wide due to the sand provided from erosion on the north end (figure 4.8).

Figure 4.5

South Bunces Key in 1990 showing that it had moved landward during the previous decade

Figure 4.6

The shoreline area where South Bunces existed and that now shows only a fairly narrow beach

Figure 4.7

Serious erosion to the northern portion of Mullet Key at the place where South Bunces Key was previously located

Casey Key

This is probably the most developed of the barriers included in this category. It is home to fairly extensive residential construction. It does not have any significant commercial enterprises. The homes are mostly on the Gulf side of the island. The island is quite secluded for this coast of Florida and is strictly for the wealthy. Its homeowners have included author Stephen King and tennis great Martina Navratilova.

The island is long and narrow, a typical wave-dominated barrier with excellent beaches (figures 4.9a and 4.9b). The coasts of this island are pristine except for the northern portion of the Gulf side. Natural wetlands extend along the landward side except for old oyster reefs that protrude into Sarasota Bay (figure 4.10). There are several of these old oyster reefs that extend perpendicular to Casey Key into Little Sarasota Bay. Their orientation is such that the reefs were at right angles to the tidal currents that moved along the bay. Oysters are filter feeders and they orient themselves so as to take advantage of the currents that carry suspended organic matter and microscopic organisms on which they feed. The oyster reefs have been buried now and support vegetation.

The Gulf side of Casey Key is mostly a beautiful beach without seawalls or *groins*. Near the north end of the island erosion problems have caused the emplacement of a *revetment* and

Figure 4.8

Very wide beach at the south end of Mullet Key benefitting from the erosion at the north end

Figure 4.9a

Looking across the island and down the beach on Casey Key showing residences

Figure 4.9b

Overview of a portion of Casey Key
showing an old oyster reef that is
extending into Sarasota Bay
from both the mainland and the
island. It has been buried by sediment
and become vegetated.

Figure 4.10

The view along the beach near the
south end of the island

walkovers (figure 4.11). For the most part, this is an island where longshore sediment transport is toward the south. There have been indications from aerial photos that at least in the north end of Casey Key sediment transport has been to the north for almost a century. This is evidenced by the migration of the unstable inlet Midnight Pass that eventually closed in the late 1980s (figure 4.12).

Manasota Key

The area near Venice is the largest coastal region along the Gulf side of the Florida peninsula where barrier islands are absent. The mainland in this vicinity includes the shore area and the Intracoastal Waterway, which was dredged largely through Miocene age limestone bedrock. Just south of this totally artificial channel, barrier islands become prominent.

The first one to the south is Manasota Key. It is a long and narrow barrier that in some respects mimics Casey Key. Where development is present it is residential. It is 11 miles long, with a nearly natural shoreline on both Gulf and backbarrier sides. A few small residential boat docks are the only structures on the landward side. The southern portion of the island is the most developed and contains some commercial units. The barrier extends from the mainland on the north to Stump Pass on the south end (figure 4.13).

The morphology of Manasota Key is typical of a wave-dominated barrier. It is long and narrow, with well-developed beaches and small dunes. The backbarrier does not show a significant amount of variability to the mangrove shoreline, indicating

Figure 4.11

Gulf side of the north end of Casey Key showing the revetment and groin structures to protect the shoreline from erosion

Figure 4.12a

Photo showing Midnight Pass migrating to the left (north) on the north end of Casey Key as it appeared in 1981, with the flood shoal covered with mangroves

Figure 4.12b

This tidal inlet has been closed since 1983.

that washover fans were not a significant part of the development of the island. Near the middle of the barrier there is a *relict* tidal inlet that has been closed, probably in the prehistoric past (figure 4.14). The inlet was closed due to a small tidal prism and the transport of sand by longshore currents moving from left to right (north to south). What was previously the north end of the southern island once had a small recurved spit on the north end that is now in a relict condition.

Figure 4.13
View of Manasota Key from the north looking south

Figure 4.14
Aerial view of a former tidal inlet that was active on the ancestral Manasota Key that has been closed due to longshore sediment transport

The beaches on Manasota Key are generally wide, with some dunes (figure 4.15a). This island has received some of the reworked sediment from the Miocene headland at Venice in the form of phosphate grains and shark's teeth (figure 4.15b). This provides further testament to the presence of a dominantly southerly transport of sediment along the beach and surf zone.

Figure 4.15a

Well-developed beach

Figure 4.15b

A close-up view of beach sediment from near the shoreline on Manasota Key showing the varied composition. The large particles are about 0.5 inches in diameter.

Gasparilla Island

This is one of the most popular barrier islands along this coast of Florida. It is primarily residential, with a small commercial area. Except for those, it is morphologically pristine. The island is a popular location for fishing, especially for tarpon in Boca Grande Inlet at the south end (figure 4.16).

Gasparilla Pass is between Gasparilla Island and Little Gasparilla Island to the north. This tidal inlet has a well-developed ebb shoal that is asymmetric, with a more linear updrift portion of the ebb shoal than the more spread-out downdrift portion (figure 4.17). In addition, the two adjacent barriers are significantly offset in the downdrift

Figure 4.16

Aerial photo of Boca Grande inlet as it currently appears

Figure 4.17

The appearance of the northern end of Gasparilla Island shown here in 1985 does not show any significant progradation caused by swash bars moving landward.

direction. This configuration is typical of what we see in the development of mixed-energy barrier morphology, but there is no evidence here for that. A comparison with the north end of Little Gasparilla Island will show this contrast.

Overall, the barrier is the typical long and narrow wave-dominated barrier.

The backbarrier is dominated by mangrove wetland and the irregular shoreline suggests that washover fans were active in the early stages of island development. Some of these former wetland areas have been developed, including a golf course. The beaches of Gasparilla Island are well developed and well used (figure 4.18). The shoreline is natural except

Figure 4.18a

a) Aerial photo and b) on-site photo of the excellent natural beach on Gasparilla Island

Figure 4.18b

for a small structure at the southern end adjacent to Boca Grande Pass. Here a seawall with *rip-rap* has been constructed (figure 4.19) to protect the lighthouse on the southern tip of the island.

Captiva Island

It is difficult to decide where to place Captiva Island in our classification of barrier islands. It is pretty continuously developed along its entire extent. The bulk of the construction is single-family and low multi-family buildings. There is no dredge-and-fill on the backbarrier area and the wetlands are still extensive there (figure 4.20).

The tidal inlets at each end of Captiva are a study in contrast. Redfish Pass on the north is tide-dominated, with a huge ebb shoal (figure 4.21). It was formed by a hurricane in 1944 and has maintained a stable position

Figure 4.19a

a) Aerial photo and b) on-site photo of the seawall that is protecting the Boca Grande lighthouse

Figure 4.19b

since that time. This has taken place because of the presence of a huge tidal prism as the result of the very large bay landward of the barrier. On the other end of the island, Blind Pass is distinctly wave-dominated (figure 4.22). In fact, it has closed multiple times and is reopened in order to maintain some circulation for water quality purposes.

Figure 4.20

Aerial photo of central Captiva Island showing the present extensive mangrove wetland on the backbarrier

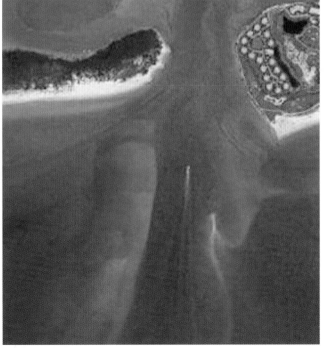

Figure 4.21

View of Redfish Pass, a tide-dominated tidal inlet that was formed by a hurricane in 1944

The presence of the large and gulfward-extending ebb shoal at Redfish Pass has prevented significant amounts of sand from bypassing the inlet. As a result Captiva Island has not received any sediment via longshore transport, only from offshore. The lack of sediment supply to the barrier caused severe erosion (figure 4.23). Over the past three decades this condition has required nourishment to maintain a beach for the heavy tourist traffic on the island (figure 4.24). Multiple projects to build the beach (figure 4.25) have been necessitated by the effects of hurricanes.

Figure 4.22

Offset of Captiva and Sanibel Islands at Blind Pass, a wave-dominated ebb shoal

Figure 4.23

Major erosion on Captiva Island with two types of groins

Figure 4.24

Nourishment construction at the same location as figure 4.23

Figure 4.25

Completed nourishment with a very wide beach on Captiva Island

Summary

This category of moderately develop-ed wave-dominated barrier islands is probably the most diverse in that we have a range of development type and even development time. Most of these islands have a significant residential component now. They are bounded by a variety of tidal inlets, but none of them causes formation of a mixed-energy morphology. The only inlet that causes changes on the adjacent island is Redfish Pass, which indirectly influences the erosion that is chronic on Captiva Island.

Chapter 5　Heavily Developed Wave-Dominated Barriers

Some of the wave-dominated barriers on Florida's peninsular Gulf coast are developed throughout with a combination of single-family and low-rise residential, high-rise condominium, and commercial buildings. They also have dredge-and-fill development on the landward side of the barrier. The general characteristics of these islands are easy access and large and dense mainland populations that give rise to development on the barriers. Most of the beaches on these barriers have been nourished; some

many times. In addition, there are hard structures along much of their shores. This discussion will consider the original characteristics of the individual islands and their current status along with some speculation on their future.

Clearwater Beach Island

One of the most famous beach areas on the Florida Gulf coast is at Clearwater. Clearwater Beach Island is developed throughout with a total spectrum of construction scales and styles (figure 5.1). This barrier has

Figure 5.1

View to the north showing the entire length of Clearwater Beach Island

no wetlands on the landward side except for the northern tip. It is all dredge-and-fill construction (figure 5.2). The Gulf side is excellent beach from one end to the other (figure 5.3).

It was not always this way. Much of this coast—about a 40-mile stretch—experienced considerable beach erosion in the late 1940s and early 1950s. Seawalls were constructed along much of it to protect the upland community. As was mentioned in the discussion in Chapter 3, the loss of seagrass in the nearshore caused a large volume of sediment to be delivered to this coast, and its presence changed the nature of Clearwater beaches. Beginning in the 1960s sand was made available and it began to naturally replace sediment on the eroded beaches. Now there is more than 100 yards of accreted sand in front of the seawalls,

much of which is vegetated and stabilized (figure 5.4). This beach and related new sand is so wide that the distance from the seawall to the shoreline could be considered a long walk (figure 5.5).

The north end of the barrier experiences a south to north longshore transport of sand. As a result, a large amount of sand moved across the mouth of Dunedin Pass and closed it in the late 1980s.

The more commercial southern portion of this island also has a really good beach. It is very heavily used and is manicured, as are many of the main tourist beaches (figure 5.6).

The south end of Clearwater Beach Island has turned a right angle corner as the result of Clearwater Pass (figure 5.7). Eventually this part of the island became developed.

Figure 5.2

Typical dredge-and-fill construction on the landward side of Clearwater Beach Island

Figure 5.3

Typical wide beach on the more private portion of north Clearwater Beach Island

Figure 5.4a

a) Aerial view of northern Clearwater Beach Island showing the wide accretion against the seawall, and b) close-up showing the seawall constructed in the early 1950s

Figure 5.4b

Figure 5.5

View down the path through the vegetated accreted sand to the shoreline

Figure 5.6

Oblique aerial photo showing the heavily used tourist beach adjacent to high-rise hotels at the south end of Clearwater Beach Island

There is north to south transport at the southern portion of the island requiring a structure to keep sand from entering Clearwater Pass (figure 5.8).

Sand Key

The longest barrier island on the Gulf Peninsula of Florida is Sand Key. It is also the most heavily developed along its entire length and in the backbarrier (figure 5.9). The dredge-and-fill in the backbarrier has eliminated virtually the entire mangrove wetland. The curvature of the barrier results in two important phenomena: 1) onshore transport diverges near the middle and moves

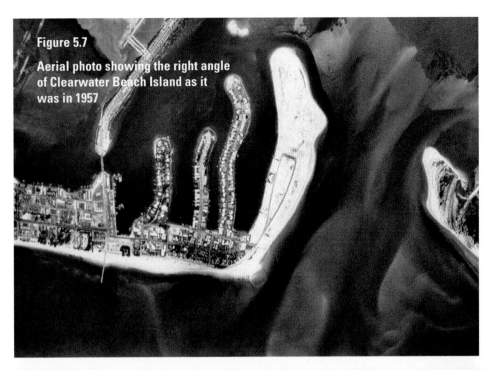

Figure 5.7

Aerial photo showing the right angle of Clearwater Beach Island as it was in 1957

Figure 5.8

Present situation at the south end of Clearwater Beach Island showing the development and a long terminal groin at the end

Figure 5.9a

Looking to the south along Sand Key at the dense development of all scales and showing the curved shoreline of the barrier

Figure 5.9b

The dense dredge-and-fill in the backbarrier

both to the north and to the south, and 2) sediment lost to erosion is not returned. The only source of sand on most of the shoreline of Sand Key is from offshore.

Erosion has been chronic for several decades and as a consequence extensive hard structures have been built (figure 5.10). The combination of the seawall and continuous buildings has prevented any washover. Fortunately there have been no major hurricanes to impact this part of the coast since extensive development has been in place. Hurricane Elena, a Category 1 storm, struck at the end of August 1985 and did cause significant damage

Figure 5.10a

a) Aerial photo showing continuous seawall and
b) close-up of typical low seawall along Sand Key

Figure 5.10b

to the beach, protection structures, and buildings (figure 5.11). It also produced the largest *ridge and runnel* feature this author has ever seen (figure 5.12). It was more than a meter high.

Because of the chronic erosion (figure 5.13), Sand Key has had a history of beach nourishment for the last 30 years. Because of the length of the island and its extensive erosion problems, the individual projects are only a few miles in length. The main problem in carrying out these projects, besides the obvious one of money, is the availability of quality borrow material—the material brought in

Figure 5.11

Damage to residences on Sand Key that was caused by Hurricane Elena in 1985

Figure 5.12

Very large ridge and runnel sediment body on Sand Key as the result of Hurricane Elena

to nourish the beach. Millions of cubic yards are required. Most of the borrow material over the years has come from the huge ebb shoal at the mouth of Tampa Bay, shown in figure 3.5 (p. 44). The material is excellent and there is almost 200 million cubic yards of it. The main problem is the distance between the borrow site and the project location, typically about 20 miles. This drives the cost per unit way up.

Each project takes months to complete and costs millions of dollars. One mode of delivery is by barge, offloading the material onto the beach where it is pumped to the desired

Figure 5.13a

Two examples of shorelines without any beach and protected by a) only a seawall and b) a combination of a seawall and rip-rap

Figure 5.13b

location (figure 5.14). The finished project is excellent (figure 5.15), but this is a temporary solution and renourishment is typically required in a few years depending on performance.

Carrying out one of these large nourishment projects requires obtaining approval from each of the local governmental units. On the northern portion of Sand Key this does not happen because one

Figure 5.14a

Beach nourishment on Sand Key showing delivery of borrow material to the project site

Figure 5.14b

Heavy equipment grading the nourished beach

community, Belleair Beach, will not give approval. All projects that utilize public funds must have public access to the beach, and this community refuses to permit anyone but residents on their beach. Consequently the beach there is not nourished, but it still receives new sand (figure 5.16) because longshore currents carry sand from the nourished beaches to Belleair Beach.

Figure 5.15a

Aerial view of completed Reddington Beach renourishment project

Figure 5.15b

Beach-level view of new, wide beach after renourishment

The tidal inlets at each end of Sand Key play an important role in the morphology of this wave-dominated barrier. The north end at Clearwater Pass was stabilized from serious northerly longshore transport of sand only 40 years ago. In its early history this inlet was called Little Pass, as seen in the coastal chart from 1883 (figure 5.17). In the early 1960s the northern tip of Sand Key was extending into Clearwater Pass (figure 5.18), causing potential problems for navigation. As a consequence plans were developed for placing a jetty structure to stabilize the inlet. Construction began in 1974 (figure 5.19a) and continued to completion with a structure that extends several hundred yards toward the Gulf (figure 5.19b).

On the other end of Sand Key is Johns Pass, a tide-dominated inlet that was formed by a hurricane in 1848. The only structure necessary here has been a terminal groin (figure 5.20).

Treasure Island

This barrier and the next one to the south are similar in character but less interesting than Sand Key. The reason is that Sand Key also has some bedrock control and that is why that part of the island protrudes

Figure 5.16

View of completed beach nourishment on Sand Key showing the narrow, non-nourished section at the community of Belleair Beach

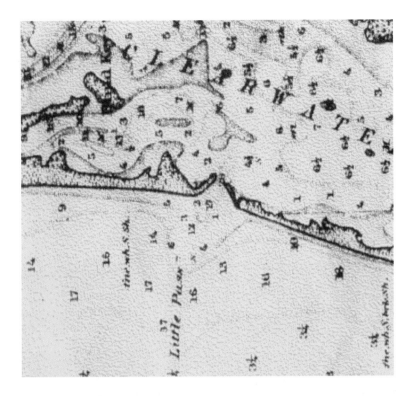

Fgure 5.17

Coastal chart showing the inlet between Clearwater Beach Island and Sand Key as it appeared in 1883 when it was called Little Pass

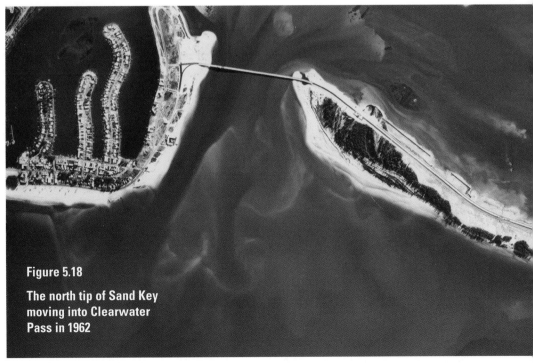

Figure 5.18

The north tip of Sand Key moving into Clearwater Pass in 1962

out into the Gulf. Treasure Island was connected to Sand Key until the hurricane of 1848 opened Johns Pass. That pass has been quite stable since its inception and Treasure Island has changed somewhat at its south end. As with Sand Key, the development of this barrier began in earnest with the construction of a causeway to it in the early 1920s. This was a time of substantial residential construction that gave way to extensive commercial development after World War II. Now the island is home to a dense combination of residences, hotels, and high-rise condominiums (figure 5.21). Prior to development and other human activities, Treasure Island was a typical narrow, wave-dominated barrier island as shown in figure 5.22, where its more recent morphology is compared with the 1883 coastal chart.

Figure 5.19a

The early stage of construction of the Clearwater jetty as it appeared in 1974

Figure 5.19b

Not long after completion of the jetty in 1979

Figure 5.20

Johns Pass, a tide-dominated inlet with a terminal groin, on the south end of Sand Key

Fugure 5.21

An overview of Treasure Island as it appeared in 1989 showing a wide nourished beach

Figure 5.22 shows that Treasure Island has quite a different appearance now compared with 1883 when it was undeveloped. It has become both wider and longer, and the wetlands have been turned into dredge-and-fill construction with numerous dead-end canals. There have been several significant changes to the nature of the island over the past several decades.

The first change was to stabilize the position of Blind Pass at the south end. The opening of Johns Pass on

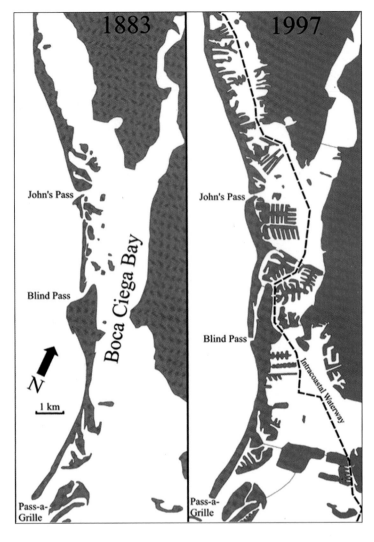

1883

1997

John's Pass

Boca Ciega Bay

Blind Pass

N

1 km

Pass-a-Grille

John's Pass

Blind Pass

Intracoastal Waterway

Pass-a-Grille

Figure 5.22

Outline maps of part of the Pinellas County coast showing southern Sand Key, Treasure Island in the middle, and Long Key at the bottom as they appeared in both 1883 and 1997 showing the changes in just over a century

the north end of the island formed a tidal inlet with a very large tidal prism. This tidal prism captured much of the volume that passed through Blind Pass, causing the inlet to become unstable and migrate to the south (figure 5.23). The inlet migrated southward about a mile until it was stabilized in 1937. As can be seen from figure 5.24, the short structure at the south end of Treasure Island permitted sediment to pass around the end into Blind Pass. The structure was extended, but the volume of longshore sand transport created a *fillet* that that still allowed sediment to pass into the inlet (figure 5.25). All of these problems with Blind Pass are the result of the small tidal prism that it carries. Maps and aerial photos have permitted a time series of outline maps to display about a century of change on Blind Pass as the relationship between Treasure Island and Long Key changed (figure 5.26).

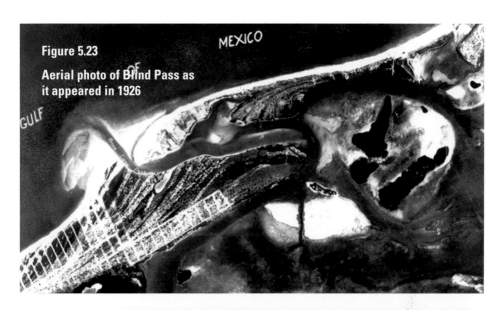

Figure 5.23

Aerial photo of Blind Pass as it appeared in 1926

MEXICO

GULF

OF

Figure 5.24

Blind Pass at it appeared in 1962 shortly after a structure was placed on the north side to stabilize the inlet

Another major change in the morphology of this barrier took place from 1966 through the 1970s. At the time there was need for a major dredging project at Johns Pass. Sediment was accumulating in that inlet because its tidal prism was reduced as the surface area of Boca Ciega Bay shrank (see figure 5.22)

by 28 percent. That means that the tidal prism was reduced by the same amount.

The amount of dredge spoil was millions of cubic yards over several years. The first dredging material was placed just offshore of the north end of Treasure Island. The material formed a small, linear lagoon that

Figure 5.25

Blind Pass looking gulfward showing a large amount of sand that has passed by the extended terminal groin and accumulated in the inlet

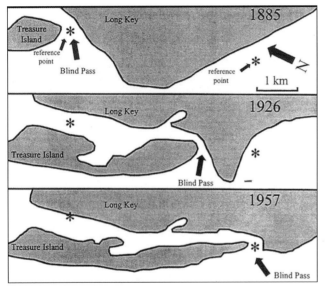

Figure 5.26

Sequence of outline maps showing the migration of Blind Pass and how it has influenced both Treasure Island and Long Key

was known as O'Brien's Lagoon, after a famous coastal engineer (figure 5.27). The water quality became an environmental problem, so a small channel, shown in the photo, was dredged to attempt to provide flushing to improve the water quality. That effort was futile. As a result, the subsequent dredging products were placed in the lagoon itself. After several of these efforts the lagoon was completely filled and the beach was groomed to provide one of the widest beaches on this coast (figure 5.28). After about three decades a new structure was added at Johns Pass on the north end of the barrier (figure 5.29). This structure is not long enough, however, to prevent sand from moving into Johns Pass from this beach.

Figure 5.27

Aerial view of O'Brien's Lagoon formed with dredge disposal as it appeared in 1974

Figure 5.28

Extremely wide beach at the north end of Treasure Island at the former site of O'Brien's Lagoon

Figure 5.29

Terminal groin constructed at the south side of Johns Pass designed to hold sand from moving into the inlet. It is too short.

Longboat Key

Longboat Key in Sarasota County is currently a wave-dominated developed barrier, but it has not always been so. The development on this island is throughout but does not include dredge-and-fill construction on the *backbarrier wetlands* (figure 5.30). A look along Longboat shows that it has a complicated development history. Near mid-island, the morphology

Figure 5.30

Oblique view of Longboat Key showing the development and the absence of dredge-and-fill in the backbarrier environment

represents a previous tidal inlet that has been closed (figure 5.31). The previously separated island had some mixed-energy characteristics as shown by the wide nature of its north end.

Longboat Key has been the site of multiple beach nourishment projects over the past few decades. Unlike most such projects, the borrow site here was an ebb shoal (figure 5.32). Permits for using this environment

Figure 5.31

View near the middle of Longboat Key showing the location of a former tidal inlet now closed and the wide end of what was a separate barrier

Figure 5.32

Dredge scar (black line) on the ebb shoal of Longboat Pass where borrow material was removed

for borrow material are typically not granted because the scars can cause changes in how waves impact the adjacent shoreline. Much of the island needed nourishment, and the final product was excellent (figure 5.33). That portion of the nourished beach that is stable for at least a year or two will become densely vegetated and will remain stable until a major storm impacts that coast. Longboat Key has been the location of multiple *hotspots*. These are places where the rate of erosion of the nourishment is abnormally high.

Lido Key

Lido Key is the smallest of the mature islands in this barrier-inlet system. It is only about 3.5 miles long and is bounded on the north by New Pass and on the south by Big Sarasota Pass (figure 5.34). The island is a major commercial and recreational area. It is distinctly wave dominated with a major north-to-south longshore sediment transport. This process causes considerable erosion to the central part of the island and multiple beach nourishment projects have taken place as a result.

The north end of the island shows a right angle shoreline at New Pass. The ebb shoal at this inlet is well developed (figure 5.35). This sediment body both protects the beach at the north end of Lido Key and also provides sediment for it. This can be seen by the welded beach ridges in the lower right corner of the photo.

Lido has been heavily developed from the time of its purchase in 1917 by John Ringling of circus fame. Under his ownership St. Armands Key was developed. A photo taken in 1945 shows limited development and then only in a portion of the island (figure 5.36.) The island now has numerous large hotels along the beach, a major shopping area on St. Armands Key (which is part of the Lido Key complex), and some single family homes. It is a very popular destination for tourists.

Figure 5.33

Excellent nourished Longboat Key beach as of 2015

Figure 5.34

Vertical aerial photograph
of Lido Key and adjacent
tidal inlets

Figure 5.35

Ebb shoal at New Pass that is providing protection
and sand for the north end of Lido Key

Figure 5.36

Aerial photo of Lido Key taken in 1945 showing only modest and local development

Lido Key has been nourished regularly and serves as a source of sediment for the beach at Siesta Key. The north-to-south longshore sediment transport carries sediment across the mouth of Big Sarasota Pass, eventually welding to the beach at Siesta. These hot spots are local coastal reaches where erosion is quite high. A major nourishment project took place in 1996 (figure 5.37a). Several years later the beach on the northern part of the island was still wide, with much of the backbeach being stabilized by planted vegetation (figure 5.37b), but other areas near the middle of the island beach needed renourishment.

The southern portion of the island is a large and pristine park. This area is extending where large amounts of nourishment sand are being transported from Lido beaches to Siesta Key by inlet bypassing.

Big Sarasota Pass on the south end of the island is essentially a one-sided mixed-energy ebb shoal. Huge amounts of sand are carried across the inlet and attached to Siesta Key to provide the wide and beautiful public beach there that has been called the nicest in Florida. It would seem that a terminal groin would help to keep this sediment on Lido Key.

Figure 5.37a

Nourished beach near the north end of Lido Key

Figure 5.37b

Grasses planted to help to stabilize the beach near the north end of Lido Key

Estero Island

The first barrier island south of the major change in position of the shoreline of this coast south of Sanibel Island is Estero Island, a completely developed barrier that is the home of Fort Myers Beach. The island is covered by a combination of residential and commercial properties (figure 5.38). As a major island for tourism, Estero's beaches must be maintained in excellent condition. Nourishment of the island's beaches has been completed, but it did not start until well into the twenty-first century. Two primary portions

Figure 5.38

Overview of part of Fort Myers Beach showing the extensive nature of the development including dredge-and-fill construction on the backbarrier wetlands

of the island were included in the nourishment program. One is the northern portion (figure 5.39) and the other is the southern two miles (figure 5.40). One problem with one of the nourishment projects is the prohibition of vegetation on a significant part of a nourished beach (figure 5.41). This is an approach that makes the beach look nice but tends to foster erosion. This part of the coast has both unvegetated (figure 5.42a) and vegetated (figure 5.42b) wide beaches. Vegetation is a great stabilizer.

The north end of Estero Island does not have a typical tidal inlet; instead, it ends in an unconfined bay. On the south end, Big Carlos Pass is a fairly dynamic inlet. A pair of photos shows major changes both in morphology and development on both sides of this feature. Prior to major development in 1958, Big Carlos was unstructured and had narrow sand bodies on both sides (figure 5.43a). By contrast a photo taken in 2015 (figure 5.43b) shows the north side had been stabilized and the sand bodies on the south side had enlarged and become vegetated, causing some degree of stabilization.

Figure 5.39

Ongoing nourishment of the northern beach of Estero Island

Figure 5.40

Nourished beach on the southern two miles of Estero Island

Figure 5.41

Large section of nourished beach that is prevented from establishing a vegetated cover to stabilize the sand

Figure 5.42a

Wide nourished beach that has no vegetation and is a favorite of tourists

Figure 5.42b

Vegetated beach that is stabilized and resists erosion

Figure 5.43a

Two stages in the development
of Big Carlos Pass: As it appeared in 1958

Figure 5.43b 2015

Summary

The densely developed wave-dominated barriers in this system range in size and orientation but have a similar appearance otherwise. The tidal inlets that border them range in size and morphology. In some places longshore transport has closed unstable inlets. Beach nourishment has caused significant changes in barrier morphology. The backbarrier wetland environments in these islands have shown tremendous change through the implementation of dredge-and-fill construction. Because these islands tend to be narrow and are highly desirable residential venues, increasing the area of development through this process makes "good use of worthless wetlands," as some would say. In fact, however, this is one of the most important environments of the barrier island system. These are among the best examples of what not to do with development of barrier islands.

Chapter 6 — Pristine Mixed-Energy Barriers

The barrier islands in the mixed-energy morphology category include all of those where one end of the island is prograding by the shoreward migration of swash bars well onto the beach. This condition depends on the ebb shoal of the tidal inlet on the updrift end of the barrier. The examples range from the transition examples of Little Gasparilla and North Captiva Islands to the classic examples of Caladesi Island and Siesta Key.

Development has changed the islands in multiple ways. In some cases the characteristic features have been masked by buildings, infrastructure, and other human modifications. Early aerial photos taken prior to development allow recognition of these features. Only Caladesi and Cayo Costa Islands remain pristine.

Caladesi Island

Probably the most classic mixed-energy barrier island on this coast is Caladesi (figure 6.1), near the north end of this barrier-inlet system. It has an excellent drumstick barrier morphology. There have been some changes over the past few decades. The only human impact has been in service of the state park: a small boat basin and buildings for rangers, a snack bar, and walkway to the beach.

Figure 6.1

Overview of Caladesi Island looking from Dunedin Pass on the south to Hurricane Pass on the north

In the early part of the twentieth century there was a small settlement on Caladesi Island that consisted of about six family residences. All of the homes were at the wide, south end of the island. The children went to school on the mainland by boat. People living on the island were essentially self-sufficient. They grew food, had a cow or two and some chickens, and they were very good fishermen and trappers. According to a book on the topic, it was a paradise.

The island was studied in some detail during the 1970s and determined to be about 3,000 years old. The current island morphology shows that a tidal inlet was present during the early development of the island (figure 6.2), when Caladesi was wave-dominated and there was an active inlet that was unstable due to longshore

Figure 6.2

Vertical view near the middle of Caladesi Island showing the relict tidal inlet that was present when this was in a wave-dominated morphology

sediment transport. Caladesi Island was formed in essentially its present state after a hurricane in 1921 that breached former Hog Island to create Hurricane Pass (see figure 1.17, page 22).

Caladesi Island has a typical wide updrift end that displays numerous beach-dune ridges (figure 6.3) formed as the result of sediment being trapped around an ebb shoal. Dunedin Pass at the south end had such a shoal that produced the swash bars (figure 6.4) that migrated landward to form these beach-dune ridges. The ridges have low areas between them that are intertidal or *subtidal*. As noted earlier, these are called catseye ponds because of their linear and narrow shape (figure 6.5).

By contrast the northern, downdrift end of the barrier is sediment starved. All of the sand that would be transported along the shoreline from the south has been trapped on the south end. The result is little

Figure 6.3

Aerial photo showing the south end of Caladesi Island where numerous beach-dune ridges have accreted onto the island

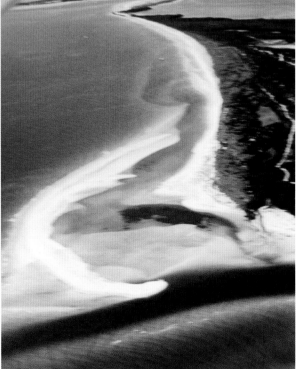

Figure 6.4

View of a swash bar just offshore of Caladesi. These are the sediment bodies that become the beach-dune ridges on the wide end of a mixed-energy barrier.

Figure 6.5

Middle portion of Caladesi showing what was previously an active tidal channel that is now closed. This closure took place before recorded history.

Figure 6.6

Shoreline of the narrow end of Caladesi where peat and mangrove remnants are exposed due to erosion and lack of replacement sand

Figure 6.7

Washover fans on the low and narrow end of Caladesi Island

Figure 6.8

In 1985, Hurricane Elena caused a breach in the narrow end of Caladesi Island, producing what was called Willy's Cut.

sediment, low elevation, no dunes, erosion, and washover (figure 6.6). The beach is very small or absent, and the shoreline is typically a peat surface with mangrove tree stumps (figure 6.7). A Category 1 hurricane in 1985, Elena, breached this end of Caladesi Island to form Willy's Cut (figure 6.8). This tidal inlet remained open for a few years but was sealed by

Figure 6.9

Overview of Caladesi showing the pristine mangrove wetland on the landward side of the barrier

sediment due to a lack of tidal prism to generate currents that would keep it open.

As a state park the island is left essentially in a natural state. No beach nourishment has been applied. The backbarrier has extensive red mangrove wetlands (figure 6.9) except some of the northern end of the island where washovers are extensive.

The south end of the barrier has experienced changes during the past few decades. The reduction in the size of the open water area landward of the island and the opening of Hurricane Pass have reduced the tidal prism for Dunedin Pass, causing it to become unstable. As a result, the inlet channel migrated to the north,

eroding the Caladesi side (figure 6.10). A storm led to the other major change related to this inlet. Hurricane Elena (1985) caused erosion of the ebb shoal at Dunedin Pass. The mouth of the inlet was being crossed by sand carried from Clearwater Beach Island by longshore transport (figure 6.11). The inlet finally closed in 1989 and has remained so since then (figure 6.12). There have been several proposals to open the inlet to permit pleasure boat traffic but none have yet been permitted.

Cayo Costa Island

Although Cayo Costa Island is a mixed-energy barrier island, it does not have the classic morphology seen in Caladesi Island. On Cayo Costa

Figure 6.10

Severe erosion on the Caladesi side of Dunedin Pass
as the channel is migrating to the north

Figure 6.11

View of the mouth of Dunedin Pass after Hurricane Elena
in 1985, which destroyed the ebb shoal and smoothed
the shoreline to permit longshore transport to the north

Figure 6.12

Dunedin Pass closed in 1989 and washover of sand from the open Gulf is common.

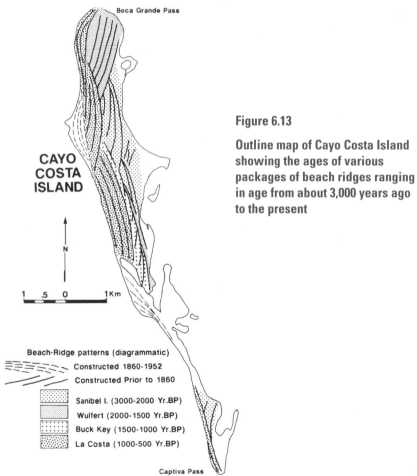

Boca Grande Pass

CAYO COSTA ISLAND

N

1 .5 0   1Km

Beach-Ridge patterns (diagrammatic)

Constructed 1860-1952
Constructed Prior to 1860

Sanibel I. (3000-2000 Yr.BP)
Wulfert (2000-1500 Yr.BP)
Buck Key (1500-1000 Yr.BP)
La Costa (1000-500 Yr.BP)

Captiva Pass

Figure 6.13

Outline map of Cayo Costa Island showing the ages of various packages of beach ridges ranging in age from about 3,000 years ago to the present

Figure 6.14

Vertical photo from 1964 of most of Cayo Costa Island showing numerous ridges in multiple packages

there are multiple places where accretion and barrier widening occur (figure 6.13). This is probably the result of complications in the nearshore bathymetry and its influence on *wave refraction*. Aerial photos taken over the past few decades have made it possible to observe the nature of the changes along the Gulf side of Cayo Costa Island.

The history of the development of Cayo Costa has been pretty well documented, with an origin about 3,000 years ago, at the time when the first barriers along this coast were formed. A detailed study of these drumstick barriers by Dr. Frank Stapor and colleagues led to the development of the map showing the chronology of the island (figure 6.14). We can see how these ridges formed by looking at a series photos taken over several years (figures 6.15 and 6.16). Intertidal sand shoals were initially formed by waves moving sand. As time passed these shoals became supratidal with vegetation, as shown in 1984 (figure 6.15b). At the present time this shoal complex has become a stable offshore barrier to Cayo Costa (figure 6.16).

Figure 6.15a

Infrared aerial photo of Cayo Costa showing multiple shoals in the nearshore Gulf

Figure 6.15b

The same area several years later (1984) showing that the shoals have become a thin and low barrier gulfward of the main island

The ridges that comprise much of Cayo Costa are only a couple of feet high and typically covered with vegetation (figure 6.17). They have accumulated in the same fashion throughout the history and development of the island. It is apparent that there has been substantial sand available with which to build this island by natural processes. At the present time there are good beaches (figure 6.18) and also some beaches where erosion is severe (figure 6.19). This is typical of islands like this as they develop. Some of the ridges become truncated as others are developing.

Figure 6.16

Cayo Costa as it appeared in 2015 showing that the small barrier has grown and appears to be stabilized

Figure 6.17

Low ridge with person standing on it adjacent to the swale without vegetation

Figure 6.18

Beach on Cayo Costa with low swash bar in the background

Figure 6.19

Erosion on the edge of the southern part of Cayo Costa Island

Tidal inlets associated with Cayo Costa Island are tide-dominated and stable. Boca Grande Pass is a wide inlet that carries a very large tidal prism. It has a typical updrift portion of the ebb shoal that extends at essentially a right angle to the shoreline and a diffuse downdrift on the southerly side (figure 6.20). As time passed it is likely that the tidal prism decreased, the inlet channel decreased somewhat in size, and the ebb shoal did also (figure 6.21). On the other end of Cayo Costa the tidal inlet, Captiva Pass, is very different in morphology. The ebb shoal is quite asymmetrical, with only the updrift (left) side present. It shows that there is considerable inlet bypassing from left (north) to right (south). It appears very similar in 1940 (figure 6.22) as it does today (figure 6.23).

Summary

There are only two pristine mixed-energy barriers along Florida's

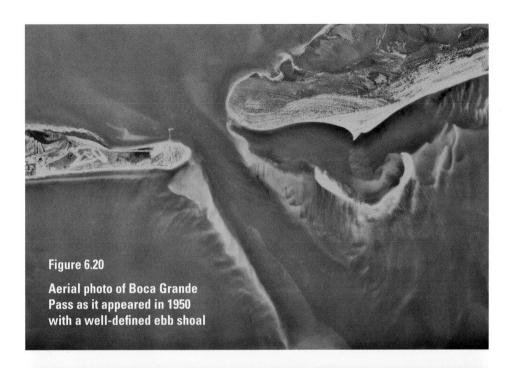

Figure 6.20

Aerial photo of Boca Grande
Pass as it appeared in 1950
with a well-defined ebb shoal

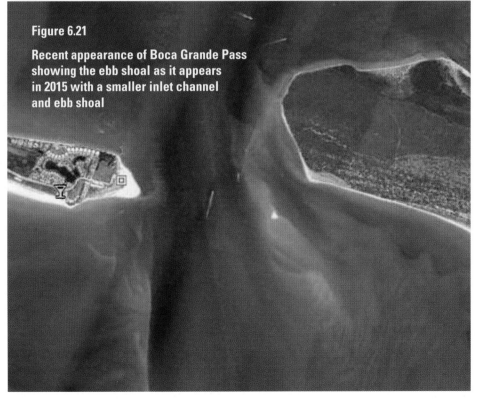

Figure 6.21

Recent appearance of Boca Grande Pass
showing the ebb shoal as it appears
in 2015 with a smaller inlet channel
and ebb shoal

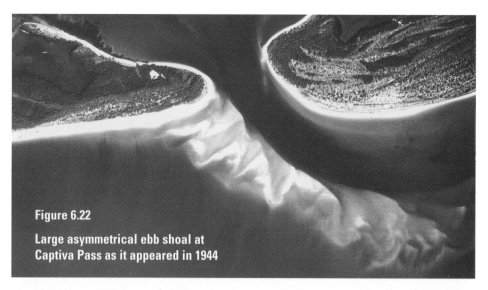

Figure 6.22

Large asymmetrical ebb shoal at Captiva Pass as it appeared in 1944

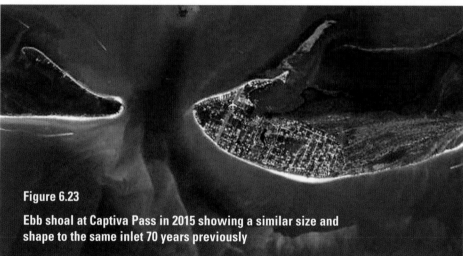

Figure 6.23

Ebb shoal at Captiva Pass in 2015 showing a similar size and shape to the same inlet 70 years previously

peninsular west coast and they are quite different in their morphology. Typically drumstick barriers on Florida's Gulf coast have been developed because their wide nature provides abundant space for a community with both commercial and residential construction. These two examples, however, are accessible only by boat and both are state parks, making human development not likely. Their natural state provides ample opportunity to study the islands and understand the nature of their origin and growth through time. Such research shows the dynamic nature of these barriers. These changes can make development both difficult and expensive and can also cause erosion problems, especially on the downdrift, narrow end of the barrier.

Chapter 7 Moderately Developed Mixed-Energy Barriers

This category comprises mixed-energy barrier islands with residential structures and widely scattered low-lying commercial structures. As in the previous chapter, there are only two barrier islands, Little Gasparilla Island and North Captiva Island, in this category. Each is in the early stages of becoming a drumstick barrier with modest progradation at the updrift end, which is north in both cases. Development on each is also in its early stages, essentially limited to residential properties on only part of the island. Both of the islands are accessible only by boat.

Little Gasparilla Island

Terminology associated with this island is quite confusing. Seven miles long, it appears on the official map of Charlotte County as "Little Gasparilla Island," so that is the name used in this discussion. The present island is a combination of multiple former islands that are named in various publications: Knight Island (formerly called Palm Island), Don Pedro Island, and Little Gasparilla Island. The name "Palm Island Archipelago" is applied to this complex along with Thornton Key, an island without any Gulf frontage. Using "Little Gasparilla

Figure 7.1

Aerial photo near the middle of the island showing single-family residences that dominate the construction of Little Gasparilla Island. Note also that there are boat docks on the back side of the barrier.

Island" simplifies the terminology and conforms to the present official map.

This island could be placed in the pristine category, but there are many residences and a couple of small inns for tourists (figure 7.1). There are about 200 homes, but only 60 people are permanent residents. There are no restaurants or stores. The backbarrier side is pristine, dense with mangroves. A single road extends essentially from one end of the island to the other. There are paths to the beach and between homes. Cars are permitted but must be carried to the island by ferry. Most visitors walk or ride a bicycle while on Little Gasparilla.

This barrier is presently in a transition stage from a wave-dominated to a mixed-energy barrier. It is long and narrow like a wave-dominated island, but the updrift, north end is developing a prograding condition as multiple beach ridges are welding on to the shoreline (figure 7.2). A series of aerial photos taken over time shows the development of both the progradation of the updrift end through multiple beach ridges being added to the barrier, and modification by human activities as roads and homes were constructed (figure 7.3). The accumulation of these beach ridges is directly the result of bypassing sand. This can be seen clearly in figures 7.3a and

Figure 7.2

Oblique aerial photo at Stump Pass showing the north end of Little Gasparilla Island with numerous accreted beach ridges

7.3c where the asymmetric ebb shoal provides the pathway from Manasota Key. The attachment of the bypassed sand is clearly shown. Figure 7.3b from 1972 shows multiple accreted beach ridges without any evidence of human interference or development. The next photo in the series, from 1985 (figure 7.3c), shows a single street platted in anticipation of home construction. Note that the attachment point of bypassed sand is apparent on this photo. The photo of the current condition of this part of the island shows numerous homes (figure 7.3d).

Some of the island has nearly continuous single-family homes on one side of the road. On part of the island, construction has produced upland home sites where boat docks are present (figure 7.4). This is the most densely developed area on the island. With the exception of a couple of homes that are too close to the shoreline (figure 7.5), a big part of the backbarrier side of the island is covered with a dense stand of red mangroves (figure 7.6).

The Gulf side shows excellent beach width and considerable vegetation to stabilize the beach, but no dunes. Another environmental positive is the setback of homes from the shoreline as evidenced by the house set back in the trees shown in figure 7.7. This is an excellent example of how stable a beach can be with good management.

The beaches on Little Gasparilla are well developed and stable. Most have been covered with vegetation to aid in this stabilization. Many parts of the island have extensive and mature plants (figure 7.8). Toward the Gulf this vegetation-covered beach is wide and flat with scattered shells debris (figure 7.9). Commonly these beaches display a ridge-and-runnel morphology as sand makes

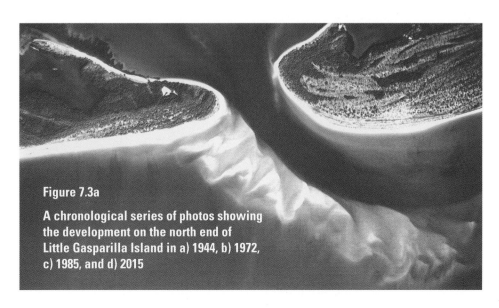

Figure 7.3a

A chronological series of photos showing the development on the north end of Little Gasparilla Island in a) 1944, b) 1972, c) 1985, and d) 2015

Figure 7.3b 1972

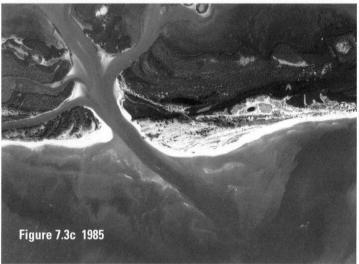

Figure 7.3c 1985

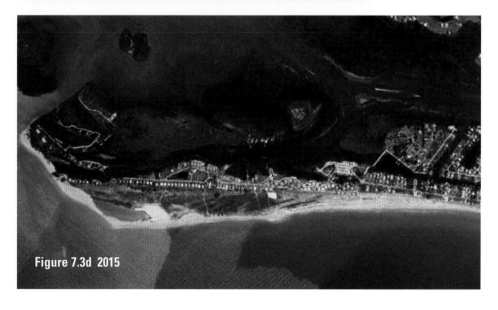

Figure 7.3d 2015

Figure 7.4

Photo showing numerous homes with boat docks
on small artificially altered water bodies on the island

Figure 7.5

Aerial photo of beach area showing homes (arrows)
that have been built too close to the shoreline

its way to the beach (figure 7.10).
Little Gasparilla beaches contain
some phosphate grains (figure 7.11),
indicating that some of the sand was
derived from Miocene strata to the
north, maybe as far as away Venice.

North Captiva Island

North Captiva is quite similar to
Little Gasparilla. It lies south of Cayo
Costa Island and has a similar initial
mixed-energy morphology to Little
Gasparilla. The north end displays

Figure 7.6

Mangrove mangal on the backbarrier side of Little Gasparilla Island

Figure 7.7

Well stabilized backbeach area with properly placed home in the woods landward of it

a complex of accreting beach ridges (figure 7.12) and the remainder of the island is narrow, with a low elevation. The island is four miles long and has a fair amount of residential construction on the wide north end. That end also includes an airstrip (figure 7.13). There is no commercial development on the island.

In order to qualify as a moderately developed mixed-energy barrier island, the updrift portion of the barrier must include a significant complex of accreted beach ridges. These ridges are mostly masked by the development on the island, but they can be revealed by a series of aerial photographs. The earliest photo

Figure 7.8

Vegetation on this beach on Little Gasparilla Island makes the beach quite stable.

Figure 7.9

Close-up of shell pavement on the beach near the south end of North Captiva Island

(figure 7.14a), taken in 1958, shows the numerous ridges that developed and moved onto the barrier before recorded history. It also shows the asymmetric ebb shoal attached to the updrift Cayo Costa Island. The photo from 24 years later shows the many streets plotted (figure 7.14b). There is no indication of any human development on North Captiva Island at this time. An infrared photo from 1982 (figure 7.15) shows the updrift

Figure 7.10

Wide and flat Little Gasparilla beach with an intertidal ridge-and-runnel typical of these Florida beaches

Figure 7.11

Low-tide beach showing dark color, indicating phosphate grains derived from Miocene strata to the north

ebb shoal and the pathway around Captiva Pass with attachment points on the island where sand accumulated after bypassing the inlet.

As figure 7.12 clearly shows, this island is not only in the early stages of developing a drumstick morphology, it is also very narrow and has a low elevation. These features make North Captiva quite vulnerable to severe storms. In 2004, Hurricane Charley hit this part of the Florida

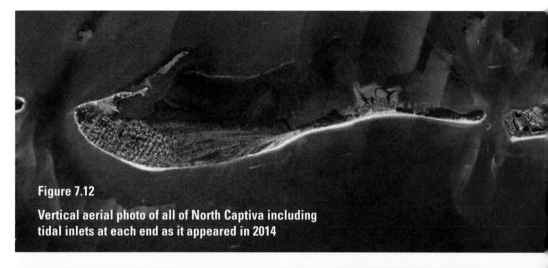

Figure 7.12

Vertical aerial photo of all of North Captiva including tidal inlets at each end as it appeared in 2014

Figure 7.13

Oblique aerial photo of the north end of North Captiva Island showing the residential development and the airstrip

coast (figure 7.16), causing serious damage, especially on North Captiva and Captiva. The nature of this damage included both the natural island environment and the built community. Charley came ashore in the middle of August as a Category 4 storm, with wind velocity of 150 mph. It caused $13 billion in damage in Florida.

The most significant change to North Captiva caused by Hurricane Charley was the formation of a huge washover fan on the low and narrow part (figure 7.17). This feature has maintained its identity until today. Charley also caused damage to the beach (figure 7.18) and to homes (figure 7.19).

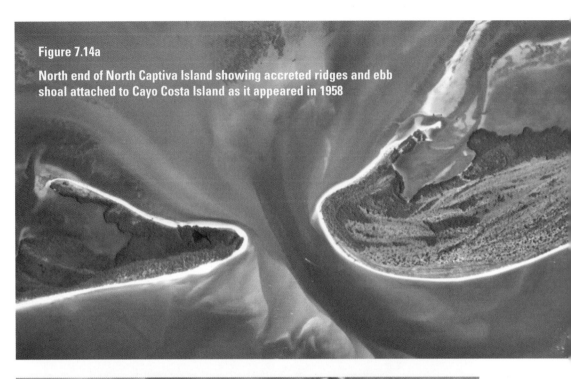

Figure 7.14a

North end of North Captiva Island showing accreted ridges and ebb shoal attached to Cayo Costa Island as it appeared in 1958

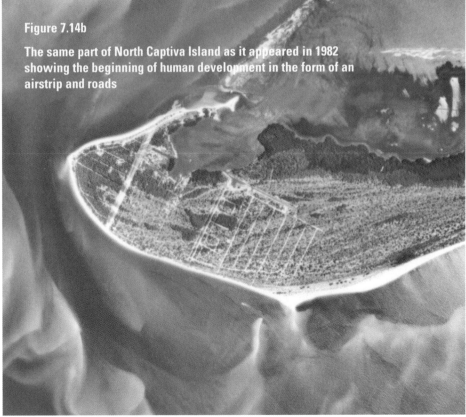

Figure 7.14b

The same part of North Captiva Island as it appeared in 1982 showing the beginning of human development in the form of an airstrip and roads

Figure 7.15

Infrared photo of North Captiva in 1982 showing the updrift ebb shoal with attachment points on North Captiva Island where sand bypassed from Cayo Costa Island reached the beach

Figure 7.16

A NOAA image of Hurricane Charley as it moved onshore of the Florida peninsula

The southern portion of North Captiva Island has been having some erosion problems for several years. The area that faces Redfish Pass has been the focus and some of the homes in this area have been threatened. Some speculation suggests that the removal of considerable nourishment material from the ebb shoal of Redfish Pass has changed the nature of wave approach to the south tip of North Captiva Island, resulting in erosion. This tidal inlet is in the tide-dominated category

Figure 7.17a

An overview of the southern half of North Captiva Island showing the large washover fan formed by Hurricane Charley

Figure 7.17b

A close-up view of the same washover fan

Figure 7.18

Trees and debris in the surf on North Captiva Island after Hurricane Charley

Figure 7.19

Severe damage to the shore area in front of coastal residences after Charley

(figure 7.21) and was formed by a hurricane in 1944 (figure 7.22a). It has a large fan-shaped flood shoal and a large ebb shoal (figure 7.22b). In order to try to mitigate the erosion problem at this end of the barrier, some nearshore structures were constructed to dissipate and minimize the wave energy that reaches this shoreline (figure 7.23).

Figure 7.20a

A narrow and steep beach on North Captiva Island

Figure 7.20b

A wide and gently sloping beach not influenced by Hurricane Charley

Figure 7.21

Overview to the south of North Captiva looking at Redfish Pass, which separates North Captiva Island from Captiva Island

Figure 7.22a

Aerial photo of Redfish Pass shortly after it was formed by a hurricane in 1944

Figure 7.22b

A modern photo of Redfish Pass showing the large ebb tidal shoal

Figure 7.23

Photo of the south end of North Captiva Island showing the structures placed to mitigate the erosion on this area. These structures withstood the impact of Hurricane Charley without significant change.

Summary

We have placed only two barrier islands in this category, North Captiva and Little Gasparilla, and they are very similar in both morphology and development. In addition they have similar tidal inlets in the updrift location relative to the barrier. Both have a similar distribution of development, which is restricted to residential buildings. The nature of these barriers is such that they are vulnerable to severe storms. An excellent example of this took place in mid-August of 2004 when Hurricane Charley struck the southern portion of North Captiva Island and caused significant damage to the island morphology and to its built environment.

Chapter 8 — Heavily Developed Mixed-Energy Barriers

Some mixed-energy barriers have become densely developed with a combination of single-family, low-rise, and high-rise residential buildings and commercial buildings. In addition, the backbarrier has been subjected to dredge-and-fill construction for many homes and finger canals. This is not an unexpected outcome given that these islands have the greatest area on which development can take place. Only those mixed-energy barriers that are not accessible by vehicle and/or are protected in some fashion have been spared heavy impact from human activities. This discussion will attempt to consider the natural morphological development of the islands, but the fact that human activities have masked the features that permit natural morphological change makes such consideration more challenging than for pristine barriers.

Long Key

Long Key is located in Pinellas County near the north end of this barrier system. Causeways connecting this barrier to the mainland were constructed in the 1920s, giving rise to human development on the island. This can be seen in the first aerial photos, taken in 1926. Figure 8.1 shows Long Key to the right (south) of Blind Pass, with numerous roads platted in a rectangle pattern. This compares with what we see

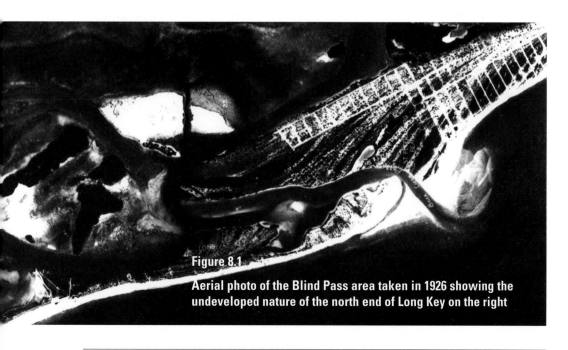

Figure 8.1

Aerial photo of the Blind Pass area taken in 1926 showing the undeveloped nature of the north end of Long Key on the right

Figure 8.2

Oblique photo of the north end of Long Key in 1979 after complete development, which took place in the 1950s and 1960s

today by looking at the same general area (figure 8.2). At the present stage of human development, the entire island is covered with buildings, roads, and infrastructure.

The early photos show us how Long Key developed into a mixed-energy barrier island. As seen in figure 8.1, numerous beach ridge features were produced as sediment was added in the form of sand bars, in keeping with the scheme for formation of a drumstick barrier (see figure 1.11, page 16). This was possible due to the well-formed ebb shoal associated with Blind Pass, the tidal inlet separating Treasure Island from Long Key.

The ebb shoal was present prior to the formation of Johns Pass at the north end of Treasure Island as a result of a hurricane in 1848. The tidal prism of Johns Pass captured much of the tidal prism of Blind Pass and caused it to be unstable. As a consequence the inlet began

to migrate with the direction of the longshore sediment transport, north to south (figure 8.3). As the inlet migrated to the south it abandoned its ebb shoal, which eventually was disbursed into the littoral sediment transport system. The migrating inlet caused erosion and truncation of the beach ridges on the north side of Long Key (figure 8.3). A look at the sequential migration of Blind Pass over more than a century shows that it moved quite rapidly until stopped by the construction of hard structures (figure 8.3).

Blind Pass continued to migrate a total of more than a mile southward. Numerous groins were placed on the southern end of Treasure Island, but sediment continued to be transported into the inlet (figure 8.4). Considerable suspensed sediment was transported into the Gulf through the inlet and longshore transport passed by the inlet to Long Key.

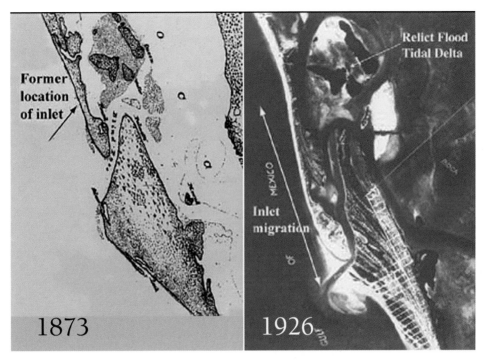

Figure 8.3

Pair of images of the Blind Pass area of north Long Key showing where the inlet was located prior to 1873 and the position of the flood shoal. By 1926 the inlet had migrated and no ebb shoal was present.

Beginning in 1937, structures were placed on the end of Treasure Island to keep sediment from accumulating in Blind Pass. The initial terminal groins were not long enough so they were lengthened. This still did not keep a large volume of sediment from moving into the inlet channel (figure 8.5).

The instability and migration of Blind Pass has been the result of two major changes to the overall morphology of this part of the coast. In addition to the opening of Johns Pass by a hurricane in 1848, the development of the barrier with extensive dredge-and-fill construction has also had

a major influence on the stability of the inlet. This construction has reduced the area of Boca Ciega Bay by 28 percent (figure 8.6). Both of these factors reduced the tidal prism greatly for Blind Pass, causing its instability. The changes caused by the dredge-and-fill construction can be compared with a pristine island of the same type to see the dramatic difference in the overall morphology (figure 8.7).

Because of its desirable location, the entire island of Long Key has been developed. It includes the town of St. Pete Beach, a large coastal community. Long Key has excellent

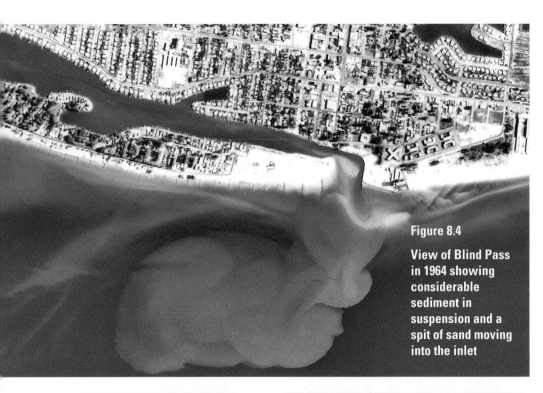

Figure 8.4

View of Blind Pass in 1964 showing considerable sediment in suspension and a spit of sand moving into the inlet

Figure 8.5

Looking across Blind Pass as it appeared in 1992 after structures were placed on each side of the inlet channel. They did not stop considerable sand from making its way through and around the structures, necessitating dredging of the inlet.

Figure 8.6

Vertical view of Long Key in 2014 showing the dense development, including widespread dredge-and-fill construction in the backbarrier region

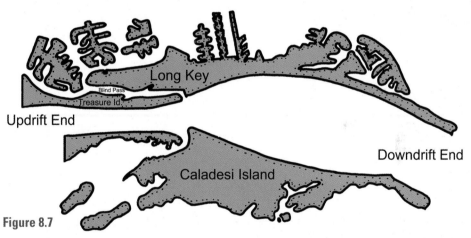

Figure 8.7

Comparison of a pristine drumstick barrier (Caladesi Island) with Long Key to show the contrast between the pristine and densely developed versions of the same barrier island morphology

beaches and public parks even though it is heavily developed. The most popular of the public beaches is Upham Beach, just downdrift (south) of Blind Pass. It is probably the most erosion-prone public beach on this coast (figure 8.8). This beach is nourished every few years as a result of the combination of erosion and sand accumulation in Blind Pass. Figure 8.8a shows this sediment and the fillet filled to capacity, with some sand passing around the end of the terminal groin on Treasure Island. Generally about 200,000–250,000 cubic yards of sand are removed and

Figure 8.8a

Upham Beach (arrows) is one of the most erosive beaches due to its down-inlet position. Here it is shown :
a) in a pre-nourishment, erosive state with abundant sand inside Blind Pass, and b) after nourishment with sand from the inlet channel.

Figure 8.8b

placed on Upham Beach. As soon as each nourishment project has been completed, people once again flock to this popular public beach (figure 8.9).

There are also places on Long Key where structures are necessary to maintain the shoreline. Sometimes a beach is present and sometimes the shoreline is against the structure, such as is shown in figure 8.10a, where the water is against the rip-rap and concrete sea wall. Further to the south on Long Key there are places where erosion is present but not severe (figure 8.10b) and others where dense vegetation has stabilized the wide beach (figure 8.10c).

Figure 8.9

Beach lovers flock to Upham Beach after nourishment

Figure 8.10a

A combination of very severe erosion with rip-rap and a concrete wall holding the shoreline

Figure 8.10b

Modest erosion on the beach

Figure 8.10c

Very stable beach with dense vegetation protecting earlier nourishment

The southernmost portion of Long Key is the community of Pass-a-Grille, a very popular beach community where nourishment has been necessary. There is a terminal groin at the tide-dominated inlet (figure 8.11).

Anna Maria Island

Anna Maria Island is the first barrier south of the main channel into Tampa Bay. It is a drumstick barrier with development throughout—a mixture of tourist facilities and permanent residences. This barrier has the typical outline of a mixed-energy barrier, but because of the intense development, no details of the wide end, such as prograding beach ridges, are visible (figure

Figure 8.11

Oblique aerial photo of the southern tip of Long Key at Pass-a-Grille showing a terminal groin and the tidal inlet

8.12). A look from north to south along Anna Maria Island shows both the density of development and the very narrow nature of the southern half of the island.

There is little historical data available to allow a detailed analysis of the origin of the morphology of this drumstick island. The overall outline of the island as it appears now supports the typical origin of such an island (figure 8.13). An aerial photo from 1941 (figure 8.14b) does not reveal multiple beach ridges as were shown in previously described drumstick barriers. The only map the author found to document this barrier's appearance prior to human development is a coastal chart from 1883 (figure 8.14a), which provides no real information on the progradation of the updrift (north) end of Anna Maria Island. My decision to place this barrier island in the mixed-energy category is primarily because of two characteristics: 1) the overall shape of the island, with a wide end and a very narrow end, and 2) the wide end being the updrift end, where abundant sediment is derived.

It is possible that this barrier is not a true drumstick island because the 1883 chart shows that it is not as wide as a typical example of this type. It actually looks more like a drumstick island in its present developed condition (see figure 8.13).

Several methods have been used to try to mitigate the erosion problems on the entire Anna Maria coast. Initially rip-rap (figure 8.15) and groins (figure 8.16) were installed, but they were not successful and they created a poor aesthetic image. More recently beach nourishment has been conducted, with considerably more success (figure 8.17).

Figure 8.12

Current aerial photo of Anna Maria Island showing the drumstick morphology and the dense development

Figure 8.13

Oblique view from north to south showing the typical morphology of a mixed-energy barrier with a very narrow downdrift end

Siesta Key

The chronology of the natural and human development of Siesta Key has been determined through multiple means. This is one of the most popular barrier islands on the Florida Gulf coast. The adjacent mainland community of Sarasota is an upscale retirement area and has been so for several decades. The main beach here has been rated as one of the best on the entire Gulf of Mexico coast. An aerial photo of the current appearance of Siesta Key shows complete human development (figure 8.18).

Siesta Key is one of the few barrier islands along this coast that has been studied in detail by Dr. Frank Stapor and colleagues in order to document

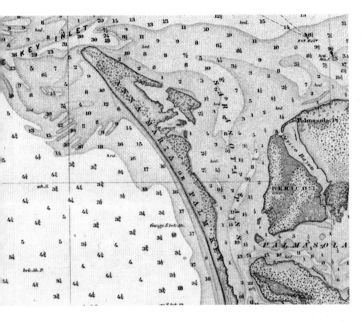

Figures 8.14a and 8.14b

a) Coastal chart from 1893 showing the outline and pristine nature of Anna Maria Island but without any details on its morphologic composition; b) vertical aerial photo of the island in 1940 prior to development

Figure 8.15

Rip-rap and seawall protecting a residence on Anna Maria before beach nourishment was undertaken on the island

Figure 8.16

Groins constructed on the southern portion of the island to attempt to hold beach sand in place. The attempt did not perform well.

its chronological development. It is one of the oldest barrier islands on this coast, even though it is only about 3,000 years old. It was then that sea level reached near its present position. When that happened, waves were able to concentrate sand into linear positions and begin the formation of barrier islands. Siesta Key is a complex of packages of beach-dune ridges that were added to a core of sand (figure 8.19a). This map contains chronological packages of ridges using the same terminology as was applied to Cayo Costa (see figure 6.13, page 120). The oldest group of ridges is called Sanibel after that island because of similar ages. Subsequent beach-ridge packages accumulated to provide the current general drumstick morphology (figure 8.19b).

Figure 8.17

A completed beach nourishment project on the northern portion of Anna Maria Island. This has proven to be the best solution to the erosion problems.

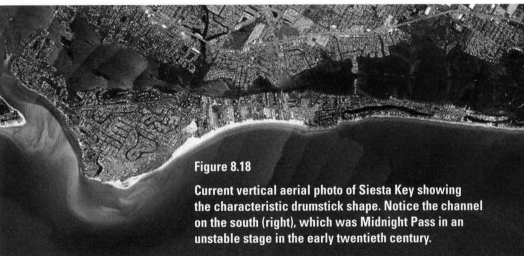

Figure 8.18

Current vertical aerial photo of Siesta Key showing the characteristic drumstick shape. Notice the channel on the south (right), which was Midnight Pass in an unstable stage in the early twentieth century.

Actual images of these ridges can be seen on early aerial photos of the north end of Siesta Key, taken prior to extensive human development. A 1948 photo shows numerous beach ridges that are not disturbed by human development but are truncated by the migration of Big Sarasota Pass (figure 8.20a). About a decade later there is some level of development in the form of roads and buildings (figure 8.20b).

The resulting appearance of the pristine Siesta Key is typical of a drumstick barrier, but its formation is a bit atypical. Typically swash bars form in the inner surf zone and move as a unit up onto the beach. Siesta Key is built by sand bypassing Big Sarasota Pass and attaching onto the

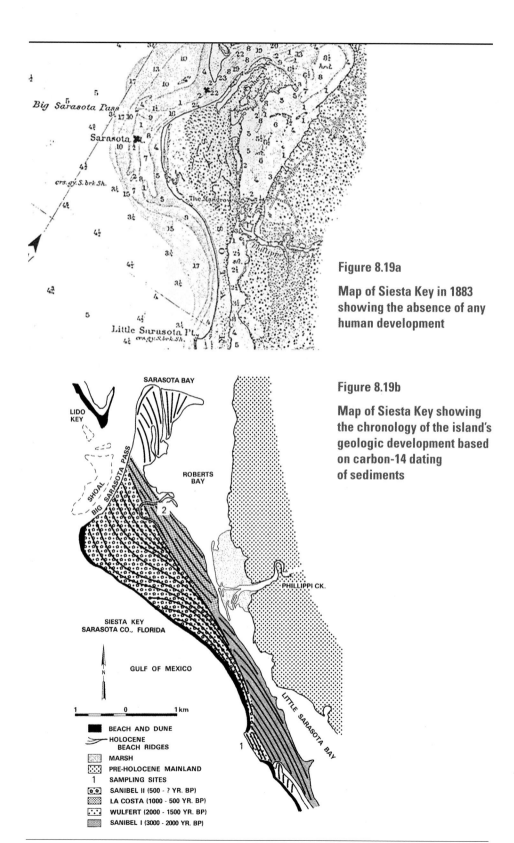

Figure 8.19a

Map of Siesta Key in 1883 showing the absence of any human development

Figure 8.19b

Map of Siesta Key showing the chronology of the island's geologic development based on carbon-14 dating of sediments

SARASOTA BAY

LIDO KEY

SHOAL

BIG SARASOTA PASS

ROBERTS BAY

PHILLIPPI CK.

SIESTA KEY
SARASOTA CO., FLORIDA

GULF OF MEXICO

N

1   0   1 km

LITTLE SARASOTA BAY

BEACH AND DUNE
HOLOCENE BEACH RIDGES
MARSH
PRE-HOLOCENE MAINLAND
1 SAMPLING SITES
SANIBEL II (500 - ? YR. BP)
LA COSTA (1000 - 500 YR. BP)
WULFERT (2000 - 1500 YR. BP)
SANIBEL I (3000 - 2000 YR. BP)

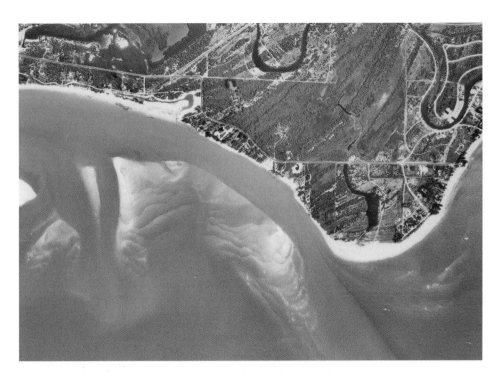

**Figure 8.20a
and
Figure 8.20b**

Two aerial photos showing numerous beach ridges on the northern portion of Siesta Key:
a) In 1948 when there was no development, and
b) in 1957 when streets and buildings were beginning to be constructed

beach, where it is distributed along the beach by wave-driven longshore currents (figure 8.21). The result is the same as on other islands, but the delivery mechanism is a bit different.

In this barrier the vast majority of the sediment that accretes to the beach is bypassed around Big Sarasota Pass and originates on Lido Key to the north.

Big Sarasota Pass has played an integral role in the morphological development and subsequent changes in Siesta Key. This tidal inlet has had a meandering path through historical time (see figure 8.18). Such a path is being forced by the southerly extension of Lido Key. A photograph from 1946 shows the entire north tip of Siesta Key being protected by hard structures (figure 8.22). Protecting the north end of Siesta Key is still a work in progress. As a consequence there has been erosion on the north end of the island, requiring the construction of structures to maintain the position of the channel margin (figure 8.23). The channel margin and northernmost part of Siesta Key are and have been an erosion problem.

Near the middle of the island on the Gulf side is a short section of the shore area known as Point of Rocks (figure 8.24a). Its name is derived from the presence of what is commonly called beachrock (figure 8.24b), a thinly bedded rock with at least about 50 percent shell hash that is deposited in the beach environment and lithifies (hardens into stone) in the surfzone. Modern artifacts such a soda bottles and beer cans are found in this type of material in the Bahamas and the Caribbean. It forms very rapidly because of the climate, sediment type, and salt water. A detailed study of Point of Rocks was conducted by Darren Spurgeon, a graduate student at the University of South Florida, to determine the age and actual formation process of this unusual rock material.

True beachrock lithfies in the beach environment in only a decade or so. By determining the chemical composition of the cement that is holding the sediment grains

Figure 8.21

A 1993 aerial photo that shows well how the sand is moving from Lido Key to the north across Big Sarasota Pass and attaches to Siesta Key.

Figure 8.22

Oblique aerial photo taken in 1946 showing numerous structures indicating major erosion problems

Figure 8.23a

Current shore protection on the north end of Siesta Key at the corner of the island

Figure 8.23b

Protection on the inlet channel wall

Figure 8.24a

Aerial photo showing the Point of Rocks coast, which protrudes out from the adjacent beach shoreline

Figure 8.2b

Close-up of the shoreline at Point of Rocks showing the beachrock

together, one can tell if it was formed in a marine beach environment or elsewhere. Spurgeon found that the lithification of the Point of Rocks beachrock took place underground in the presence of fresh water. It is not, therefore, true beachrock. It is composed of beach sediment, but lithification did not take place in the beach environment. He also found by carbon-14 dating that the rock is older by about 1,000 years than the oldest part of the current Siesta Key. The interpretation is that this material was part of an older barrier island that became buried. Groundwater then percolated through it, precipitating calcium carbonate cement.

On the landward side of Siesta Key near this part of the island there are small land bodies that extend into Little Sarasota Bay (figure 8.25). These are relict oyster reefs that have been covered with sediment, some of which was derived by the dredging of the Intracoastal Waterway.

The final part of the discussion of Siesta Key deals with the historical changes that have taken place in the tidal channel of Midnight Pass. This inlet developed a very large flood

Figure 8.25

View looking landward across Siesta Key

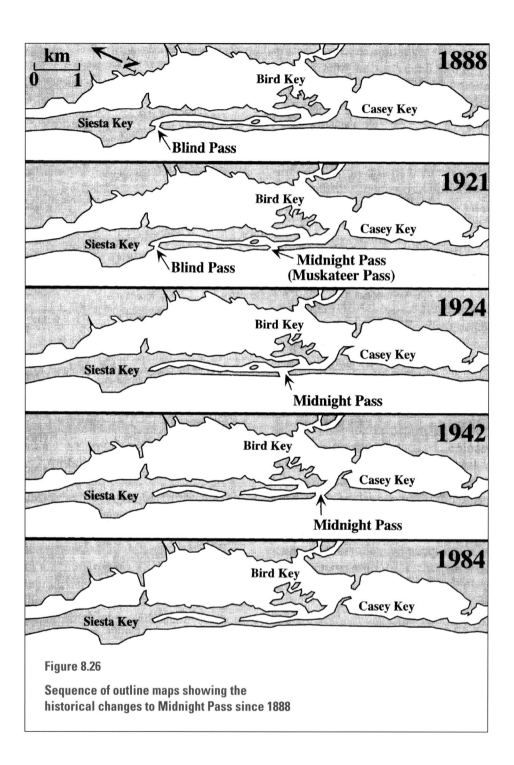

Figure 8.26

Sequence of outline maps showing the
historical changes to Midnight Pass since 1888

Figure 8.27a

The state of Midnight Pass as it looked in 1983 just before closure

Figure 8.27b

Midnight Pass after closure, as it still remains

shoal, but a small ebb shoal. The first fairly accurate map shows the inlet channel has migrated about 2 miles (3 km) to the north (left) and was stopped from further movement by the presence of Point of Rocks (figure 8.26). The large relict flood shoal can be seen in Little Sarasota Bay landward of the barrier. The long spit produced by this migration has been breached multiple times. Eventually by the early 1980s the inlet channel showed signs of closing. Finally it did so in 1984 and has remained closed since that time (figure 8.27b).

Summary

The drumstick mixed-energy barriers discussed in this chapter are not all of the same origin. Because they are densely and diversely developed by human activity, it is difficult to obtain information about them prior to that development. Although their overall shape is similar and in keeping with drumstick configuration, they were formed by somewhat different mechanisms. The basic formation did, however, concentrate sand at one end at the expense of the other. The adjacent tidal inlets are very important in the morphodynamics of these islands and can cause problems after development if the inlet becomes unstable.

Chapter 9 Unusual Barriers

There are a few barrier islands along the Florida peninsula that do not fit into the categories covered in the previous chapters. Their morphology does not conform well with either the wave-dominated or mixed-energy categories. Two of them might be classified as mixed-energy, because they show accretion of beach ridges, but they do so in a different fashion. Each also has a unique story of growth and development. Three of the four of these barriers are here considered to be moderately developed, but for quite different reasons. Honeymoon Island had its morphology greatly modified by a residential developer in the 1960s, resulting in serious problems ever since. Egmont Key had a town and a military compound more than 100 years ago. Sanibel Island has experienced residential and commercial development in the traditional way. Development on Sanibel has occurred mostly along the shore area, leaving most of the rest of the island in pristine condition, including considerable wetlands of mangrove mangals.

Honeymoon Island

Honeymoon Island State Park is currently one of the most popular parks in the Florida system. It is located across Hurricane Pass from Caladesi Island and is accessed by the Dunedin Causeway near the north end of this barrier-inlet system. Prior to a hurricane in 1921, these two islands were joined into one, called Hog Island (figure 9.1). The tidal inlet formed by the hurricane was viable from its early stage (figure 9.2a) and remains so to the present (figure 9.2b). This island was very narrow on the south end and was becoming wider at the north end. The former Hog Island was unusual in that it had a curved shoreline that was related to the offshore bathymetry. As a result, as waves approached they were refracted so that longshore currents flowed in both directions: to the north on Honeymoon and to the south on Caladesi. As a consequence the sand carried along the beach on Honeymoon was being transported to the north and accumulated as shoreward migrating swash bars. As time passed this northern end of the island became long and relatively wide (figure 9.3).

The Dunedin Causeway, which connects the mainland with Honeymoon, was constructed in 1963–64 and paved the way for considerable visitation to the island and its coast. Prior to that time the island was visited only by boat and was a common site for weddings, hence the name. In the later 1960s the island was transformed in a way that has caused problems to this day.

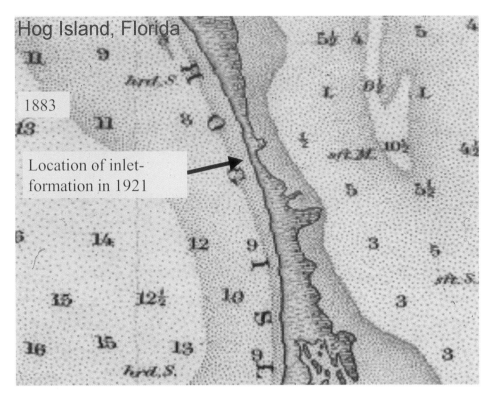

Figure 9.1

Coast and geodetic survey map of Hog Island from 1883 showing the location of Hurricane Pass, with Honeymoon Island on the top and Caladesi Island on the bottom

A private corporation owned the island and in its greed decided to plat an entire community there. The only problem was that the island didn't have enough land for the plan, so the corporation set about enlarging it. Fill material was dredged from about 350 yards offshore and placed along the shoreline to enlarge the island and raise the beach about 3 feet above sea level (figure 9.4). Prior to the project there was a detached spit on the north portion of the island and after completion the shoreline of Honeymoon Island was directly connected to that spit. The dredged

borrow material was a combination of sorted sand and limestone boulders and cobbles from the shallow Tertiary bedrock. This was an ill-fated project in that most of the 1,140,000 cubic yards of borrow material was the limestone debris. In a short time the sand had washed away and the beach was all rip-rap (figure 9.5). The state purchased Honeymoon Island in the early 1970s as parkland. The existing conditions were a major problem for a beach-centered state park.

The sand that had been placed along the Honeymoon shoreline covering the rip-rap limestone was removed

Figure 9.2a

Vertical aerial photo of Hurricane Pass
a) shortly after its formation (1926) showing
a large flood shoal and b) to this day still
with a flood shoal

GULF OF MEXICO

HOG   ISLAND

Figure 9.2b

and transported in both directions by longshore currents. This sediment supply extended the island toward Hurricane Pass to the south (figure 9.6a) and also formed a spit to the north of the island (figure 9.6b).

Because of the popularity of the park and the serious problems with the loss of sand and all of the rip-rap on the shoreline, a nourishment project was planned in the late 1980s. In a very unusual approach, the decision was made to use an upland source

Figure 9.3

The spit on the north end of Honeymoon Island has grown both long and wide over time.

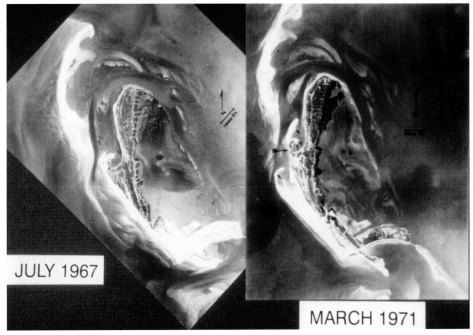

JULY 1967

MARCH 1971

Figure 9.4

Comparison of aerial photos of Honeymoon Island in 1967 prior to the nourishment and widening of the island and in 1971 just after the completion of the enlarging of the island

Figure 9.5

Rip-rap at the shoreline after all of the sand has been removed from the central part of the island

Figure 9.6a

Sand removed from the central portion of Honeymoon Island was a) transported south along Hurricane Pass, and b) carried to the north forming a spit with a well-developed beach.

Figure 9.6b

as borrow material for nourishment. *Pleistocene* dunes near Tarpon Springs were used for the project. The dune sand was hauled by dump truck to the project site on the Honeymoon Island beach. Just under 250,000 cubic yards of material were placed on the central beach, the location most visitors preferred.

This project basically returned the shoreline to its position after the widening of the island in the early 1970s. There were multiple problems, however. First, the roads were damaged by the transport of thousands of truckloads of sand. Although most of the trucking was done late at night, it still caused traffic problems. Visitors expressed immediate displeasure with the color of the new sand (figure 9.7), which was stained a reddish brown, typical of most inland dunes in the area. The cause was a thin film of iron oxide that coated the quartz grains.

The worst aspect of the project was that the sand did not last very long where it was placed. In only about a year the sand moved primarily to the south and into Hurricane Pass. Because of the color of the nourishment material it was easy to follow its path to the north and south by longshore currents. Erosion was prominent within the first year (figure 9.8).

Honeymoon Island was not in good shape for visitors until after the turn of the twenty-first century when a series of efforts were undertaken to fix the problem. Another nourishment project included a terminal groin to keep the sand from moving south and into Hurricane Pass (figure 9.9). This solution did not last very long and erosion became prominent again (figure 9.10). Efforts to solve these problems continue to the present time, including building hard structures and additional nourishment.

Figure 9.7a

Colored sand from the nourishment project on the beach at Honeymoon
State Park a) as it appeared in 1989 at the nourishment site and b) two years
later as the nourishment sand had made its way into Hurricane Pass

Figure 9.7b

Figure 9.8a

Severe erosion and exposed rip-rap on Honeymoon Island from a) the air and b) at the shoreline

Figure 9.8b

Figure 9.9

Terminal groin as part of a more recent nourishment project in 2007

Figure 9.10

Erosion in 2015 on Honeymoon Island, with rubble on the beach just before nourishment

Egmont Key

Egmont Key is located across the mouth of Tampa Bay (figure 9.11) and is an unusual island. It is not a true barrier because of its location on the huge ebb shoal at the mouth of Tampa Bay. It is elongate and oriented with the other barriers, however, and it is important enough to be included in this discussion. Except for an abandoned fort and a base for the Tampa Bay pilot's association, it is pristine at this time, although it was developed in the past.

This island originated from wave action much like the others discussed earlier, but the sediment came from the huge accumulation at the mouth of Tampa Bay. This sediment accumulation amounts to more than 200 million cubic yards of sand and is the second largest sediment body that extends beyond the primary shoreline into the Gulf of Mexico. The largest is obviously the Mississippi River Delta in Louisiana.

The sediment body is essentially equivalent to the tidal shoals that

Figure 9.11

Aerial view of Egmont Key showing extensive sand shoals that extend toward the Gulf of Mexico

form adjacent to most tidal inlets along this coast. The location is positioned just seaward of the channel and is called the ebb tidal shoal. This huge sediment body was formed as tidal processes became dominant during the rise in sea level that occurred after glaciers melted. Initially the rapid rate of sea-level rise caused tidal processes to dominate the coastline. About 7,000 years ago that rate of rise slowed and waves began to be prominent in coastal morphology. Another slowdown occurred about 3,000 years ago when the Florida barriers began to form. This island was formed by wave action and remains wave-dominated. The northern end of the island terminates at the main channel that extends out from Tampa Bay. It is almost 100 feet deep in this area and is maintained at that depth by natural tidal currents.

This provides an excellent pathway for the busy ship traffic in and out of the Port of Tampa.

Egmont Key has an unusual history. It was home to Fort Dade, a military compound that was constructed in the late nineteenth century along with Fort De Soto on adjacent Mullet Key. The fort was adjacent to a town (figure 9.12), the only remnants of which are a small portion of the light rail track (figure 9.13) that was used to transport heavy guns across the island. The overall appearance of the island has changed markedly (figure 9.14). The structures associated with the fort are largely still standing, but in a quite different preservation state (figure 9.15). Erosion has badly damaged some of them (figure 9.16). Erosion problems in the north portion of the island go back to the early 1900s when the fort was

Figure 9.12

View of the town on Egmont in the early twentieth century when the fort was occupied

Figure 9.13

Close-up of the remaining portion of the light rail line on the island

Figure 9.14

Overview showing Egmont Key now covered with trees

Figure 9.15

An example of the heavy mortars that were part of Fort Dade

Figure 9.16

Part of the fort on Egmont Key crumbling into the beach and shallow nearshore

new. The first attempt at preserving the batteries was undertaken by the Corps of Engineers, which constructed groins and a seawall, but the effort was futile.

The morphological history of the island is well documented for more than a century based on maps and aerial photos. The main trend over this period has been erosion, especially on the southern portion (figure 9.17). Virtually all of the erosion has been on the Gulf side of the island (figure 9.18). This is expected because it is the side where waves impact the shoreline.

The present condition of the beach environment on Egmont Key is varied; in some places the beach is severely eroding and in other places it is accreting (figure 9.19). In general the west side facing the Gulf is losing beach and upland material, forming small bluffs and leaving mature trees on the beach's *swash zone*. It is evident that much of the sand being removed is moving along the beach to the south where it is accumulating (figure 9.20). Overall erosion in the early part of this century has reduced the area of Egmont Key by about one-third.

Preservation of portions of the historic fort is a priority, and attempts have been made to do so. Beach nourishment and *geotextile groins* were placed on critically eroding portions of the west side of the island in 2002 and 2006. This

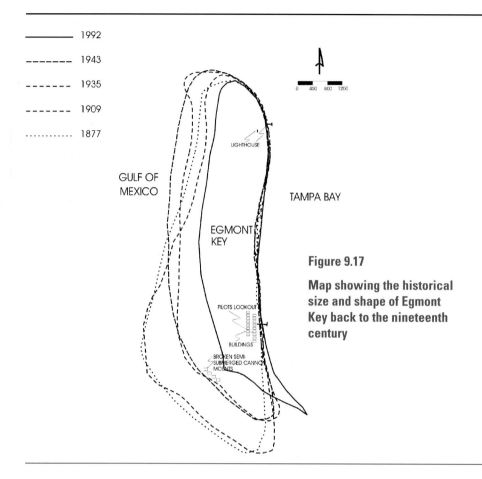

1992
1943
1935
1909
1877

GULF OF
MEXICO

LIGHTHOUSE

TAMPA BAY

EGMONT
KEY

PILOTS LOOKOUT

BUILDINGS

BROKEN SEMI-
SUBMERGED CANNON
MOUNTS

0   400 800 1200

Figure 9.17

Map showing the historical size and shape of Egmont Key back to the nineteenth century

approach has been fairly successful. A nourishment berm of 50,000 cubic yards of sand was placed offshore, but it was undetected shortly thereafter.

Egmont Key is an unusual entry in this category of barrier islands, but it looks and behaves somewhat like the others even though its origin is quite different. It is exposed to open Gulf processes, and a significant hurricane could have major impacts on it.

Sanibel Island

Sanibel is the most unusually shaped barrier island in this system. It extends south from Captiva Island,

looking much like a typical barrier, but then it turns the corner and the remainder of it is oriented essentially west-east (figure 9.21). This change in shoreline orientation is most likely due to a change in the contours of the underlying bedrock. Previous authors who investigated the geology of this area found that only about 20 feet of sand rests on top of the older *stratigraphic* units that are somewhat lithified. It is this surface that turns the corner and causes Sanibel to do the same.

It is obvious that Sanibel Island was formed by the accumulation and progradation of beach ridges against

Figure 9.18a

Erosional beach on Egmont that has moved landward of mature trees showing a significant loss of beach sand on the Gulf side

Figure 9.18b

Egmont Beach showing an erosional scarp, also on the Gulf side

the relict shoreline as the island experienced its natural development. It is the way in which these beach ridges attach to Sanibel that has put the island in this category. A look at the barrier prior to significant coastal human development shows the orientation of the welded beach ridges to be at various angles. A detailed study of these ridges by Dr. Frank Stapor has shown the island to be about 3,000 years old, and several packages of beach ridges are on the barrier (figure 9.22). There is no situation on Sanibel or the adjacent Blind Pass where an ebb shoal could

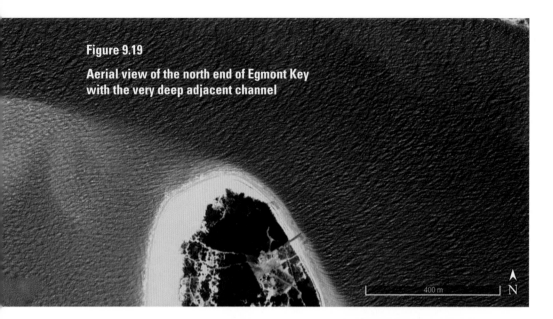

Figure 9.19

Aerial view of the north end of Egmont Key with the very deep adjacent channel

400 m

N

Figure 9.20

Aerial view of the south end of Egmont Key showing the large amount of accretion on the beach

be present to cause formation of a mixed-energy barrier. The sand that made these ridges and formed much of the island simply came to it through longshore sediment transport. This can be seen by the ridges that comprise the outer few elements on the barrier as it exists today (figure 9.23).

An overview of Sanibel Island shows its large size and also that much of the island is covered with mangrove mangal on the landward

Figure 9.21

Current aerial photo of Sanibel Island showing the change in orientation as the shoreline extends from the north

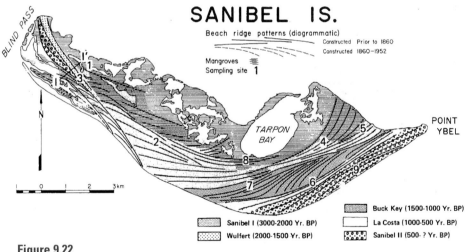

SANIBEL IS.

Beach ridge patterns (diagrammatic)

———— Constructed Prior to 1860

– – – – Constructed 1860–1952

Mangroves ≣

Sampling site **1**

TARPON BAY

POINT YBEL

BLIND PASS

N

0   1   2   3 km

Sanibel I (3000-2000 Yr. BP)

Wulfert (2000-1500 Yr. BP)

Buck Key (1500-1000 Yr. BP)

La Costa (1000-500 Yr. BP)

Sanibel II (500- ? Yr. BP)

Figure 9.22

Map of Sanibel Island showing the various packages of beach ridges and their ages

side (figure 9.24). The substrate on this wetland community is primarily from washover deposits. Human development and Sanibel's famous beaches are concentrated on the other side (figure 9.25). Unlike most of Captiva Island to the north, Sanibel has had little nourishment. The beaches are in good shape (figures 9.26, 9.27). Shells are abundant and many people visit the island to collect them.

Figure 9.23

Aerial photo showing the wide beach ridges with large catseye ponds near the bend in Sanibel

Figure 9.24

Photo across Sanibel showing the concentration of development near the shoreline and the extensive mangrove community on the landward side

Marco Island

Marco Island is probably the most densely developed barrier on the Florida Gulf coast. It is short and wide and all of it is under development (figure 9.28). The island is a complex of high-rise and single-family residences with some commercial buildings scattered throughout. Like many of these barriers, especially on the south Florida Gulf coast, development began in the 1960s.

A first glance at Marco Island suggests that it must be a mixed-energy barrier because of its width. It really does not have the proper morphology, however, because it is quite wide throughout and both ends

Figure 9.25

Oblique aerial showing Sanibel Island where it "turns the corner"

Figure 9.26

View of one of the popular beaches on Sanibel Island

of the island protrude into the Gulf. The present condition of the island does not permit recognition of any aspects of its natural morphology. Fortunately some pre-development aerial photos are available that permit us to get a pretty good interpretation of the natural development of the island. A photo taken around 1960 shows essentially no development (figure 9.29). In its early history the island was settled by Calusa Indians for hundreds of years. A wealthy northerner, Baron Collier, bought the property and began a small settlement in the northeast corner in 1870. The railroad came in 1912 and the first bridge allowing automobile traffic to the island was completed in 1938.

Figure 9.27a

Beach view of one of Sanibel's outstanding beaches

Figure 9.27b

Probably the most famous beach on Sanibel, near the lighthouse

Based on the earliest photos it is apparent that the sand portion of Marco Island came from a combination of longshore sediment transport and onshore transport. No significant ebb shoal around which wave refraction and sediment transport reversal could take place was present. This looks like a wave-dominated barrier, but it is very wide and it does not have the characteristics of a wave-dominated barrier.

Figure 9.28

Vertical aerial photo of Marco Island showing total human development throughout the island

Figure 9.29

Historical overview of the Gulf side of Marco Island before development at the end of the 1950s

Figure 9.30

The initial development recorded showing the beginnings of roads and buildings

The first significant development was in the early 1960s when some roads and small buildings were constructed (figure 9.30). The rate of development increased rapidly through the sixties. Extensive activity throughout the island and into the wetlands (figures 9.31a and 9.31b) on the landward side eventually led to complete human occupation as it is now (figure 9.32).

With extensive development can come some problems. Some offshore breakwaters were installed at the south end and nourishment for the beach was required along the northern part of the island. This project was both large and successful (figure 9.33). At the present time there is also natural sand accumulating as a spit the north end of the island (figure 9.34).

The current situation on the beaches of Marco Island is good. They are wide and they protect the upland environment well (figure 9.35). Along most of the shoreline there is a low seawall between the active beach and the upland. The south end of Marco Island does have minor erosion problems requiring construction of protective structures (figure 9.36). Caxambas Pass is not a completely stable pass, but it is not a major problem. The structures are holding it in place well.

Figure 9.31a

Major development including dredge-and-fill construction in 1964 by the Deltona Corporation, the major developer of Marco Island

Figure 9.31b

Photo of development looking toward the Gulf as it proceeds toward completion

Figure 9.32a

The completed development of Marco Island is shown a) from a vertical view showing different elements of the island and b) looking toward the island from the Gulf

KEEWAYDIN ISLAND

CAPRI PASS

HIDEAWAY BEACH

BIG MARCO RIVER

TIGERTAIL BEACH

CENTRAL MARCO BEACH

MARCO ISLAND

GULF OF MEXICO

SOUTH MARCO BEACH

CAXAMBAS PASS

Fibure 9.32b

Figure 9.33a

A view of the northern part of Marco beach before nourishment

Figure 9.33b

The same view of the excellent beach after completion of the project

Figure 9.34

The spit at the north end of Marco as it appears now shows how the sediment is moving along the shore.

Summary

With so many barrier islands along this coast, it is expected that a few would not fall into our rather simple classification scheme. The four examples discussed here are a rather broad spectrum of situations regarding location, underlying influence, and human impact. They each have different histories but

Figure 9.35a

The nourished beach on the north part of the Marco Island

Figure 9.35b

The small stabilizing structure on the southernmost beach

Figure 9.36

A view of Caxambas Pass at the southern end of Marco Island

their morphology can be explained. They are all the result of wave and tidal processes but those processes acted in different ways. Two are state parks and the other two are among the most popular destinations on the Florida Gulf coast.

Conclusions

As was mentioned in the beginning of the book, the barrier island system of Florida's Gulf coast is the most morphologically complicated in the world. By complicated I mean that the different shapes of barrier islands and tidal inlets cover the complete spectrum of the known types. There are other aspects of this barrier-inlet system that make it both complicated and unique. The influence of the underlying bedrock topography on the position of the barriers is a very important part of the system. Another critical factor is the size of the bays. On top of all of these natural conditions on this coast is the impact of development caused by humans.

The complicated morphology of the barrier system began before the barrier islands formed. As sea level rose from the beginning of post-glacial time, the coast was tide-dominated. Sea-level rise slowed down substantially at about 7,000 years before present and waves began to modify the coast. From the time the barrier islands and inlets began to form about 3,000 years ago until about the early 1920s, the system was pristine, with no significant human intervention. The most important factors of change in those early years of the barriers were severe storms. The ones that we know about were

the hurricanes of 1848 and 1921. The one in 1848 cut Johns Pass and the one in 1921 cut Hurricane Pass, both in Pinellas County. Further to the south, Redfish Pass was cut in 1944. All three of these have persisted since their formation, with Johns Pass and Redfish Pass being tide-dominated and Hurricane Pass being moderately wave-dominated.

The big change came after construction of causeways from the mainland began in the 1920s. These were fill causeways that not only permitted unrestricted vehicular traffic but also caused some changes in tidal circulation within the bays and through the inlets. This was just after World War I and it was the beginning of significant tourism from the North, especially during the winter. With few exceptions, residential and commercial buildings sprang up on the barrier islands. Tidal inlets were stabilized and beaches were being protected to help stop erosion.

The next step in the development of the barrier system was the dredging of small ditches in the wetlands on the back of the barriers in an effort to control mosquitoes that are very common in this environment (figure C.1). The objective was to dry the salt marshes landward of the barriers and thereby remove the habitat of the

Figure C.1

An example of a mosquito ditch, the type that was dug in rectangular patterns on the salt marshes landward of the barrier islands

Figure C.2

Dredge-and-fill construction on the landward side of a barrier island near Clearwater. A fill causeway is shown in the background.

mosquito. It did not work, but it did harm the marshes by removing some of the tidal flux that they depend on for sustenance. This part of the barrier system complex was greatly modified shortly thereafter through the practice of dredge-and-fill construction. Canals were dredged through the salt marsh and mangrove environments with the dredge spoil being placed adjacent to the canals, providing an upland location for

Figure C.3a

Shore-parallel structures such as a) seawalls and b) breakwaters have been constructed in many places.

Figure C.3b

housing (figure C.2). This procedure has caused major problems with the barrier island environment and has been illegal since the late 1960s.

Open water structures on the barrier-inlet coast have been built for about a century. These structures are designed to provide protection for the open beach coast and for tidal inlets. Erosion on the beaches caused the need for seawalls, breakwaters, and groins. Seawalls are shore-parallel and have not been very successful in any way except protecting the upland property. Breakwaters are offshore structures that have had mixed results (figure C.3). Groins are shore perpendicular and are designed to stabilize the beach and longshore transport of sand (figure C.4).

Over the past few decades these approaches to beach management have changed to nourishment. Sand is dredged from an appropriate source and placed on the beach, then molded to the design specifications

Figure C.4a

Generally small shore-normal structures include a) geotextile groins and b) hard fence-like groins.

Figure C.4b

Figure C.5a

The initial effort in beach nourishment is dredging from an approved borrow site and pumping the product to the beach.

Figure C.5b

At the beach, the material is molded into the design profile.

(figure C.5). Such soft protection is temporary and expensive, but it has become the norm because it provides an excellent beach for both recreation and protection of the upland structures (figure C.6).

Because most tidal inlets are at least somewhat unstable and/or they receive sand that moves along the coast as longshore drift, they are structured. These structures are in

Figure C.6a
Aerial view of a recently nourished beach on Sand Key

Figure C.6b

A close-up view of a recently nourished beach at Reddington Beach on Sand Key

the form of jetties to stabilize the channel for navigation and terminal groins that are designed to prevent longshore transport of sediment from entering the inlet (figure C.7). The terminal groins eliminate the need for channel dredging in the inlets if they are long enough to prevent sediment from passing the end of the structure.

Figure C.7a

Two examples of common tidal inlet structures are a) a terminal groin on the south side of Johns Pass and b) the jetties at Venice Inlet prior to nourishment on the south (right) side.

Figure C.7b

To summarize the present situation on this important and extensive Florida barrier-inlet system, it can be evaluated only as fair. Erosion continues, many structures are not performing well, and many other problems exist. Obviously development has proceeded quite far and much detrimental activity has taken place. Fortunately, some barrier islands remain pristine or nearly so. It is important to manage these properly. One hopes that we have learned something from the mistakes of past development activities that will serve well in the future as coastal management proceeds.

Important Readings

Brooks, G.R. (ed.). "Neogene Geology of a Linked Coastal/Inner Shelf System: West-Central Florida." *Marine Geology*, v. 200, 2003. (Seven of the papers in this volume are on this barrier island system.)

Davis, R.A. (ed.). *Coastal Sedimentary Environments*, 2nd ed. Heidelberg: Springer-Verlag, 1958.

Davis, R.A. 1988, "Morphodynamics of the West-Central Florida Barrier Islands System: The Delicate Balance Between Wave and Tide Domination." *Coastal Lowlands Geology and Geotechnology*, pp. 225-235. Dordrecht, Netherlands, 1988.

Davis, R.A. (ed.). *Holocene Barrier Island Systems*. Heidelberg: Springer-Verlag, 1994.

Davis, R.A. "Barrier Island Systems— Geologic Overview," in Davis, R.A. (ed.), *Holocene Barrier Island Systems*, pp. 1-46.

Davis, R.A. "Barriers of the Florida Gulf Peninsula," in Davis, R. A., (ed.), *Holocene Barrier Island Systems*, pp. 167-206.

Davis, R.A. *Beaches of the Gulf Coast.* College Station, TX: Texas A&M University Press, 2014.

Gibeaut, J.C., and R.A. Davis. "Computer Simulation Modeling of Ebb-Tidal Deltas." *Coastal Sediments* '91, pp. 1389-1403. Seattle: American Society of Civil Engineers, 1991.

Hayes, M.O. 1979, "Barrier Island Morphology as a Function of Tide and Wave Regime," in Leatherman, S.P. (ed.), *Barrier Islands*, pp. 1-29. New York: Academic Press, 1979.

Hine, A.C., M.W. Evans, R.A. Davis, and D.F. Belknap. 1987, "Depositional Response to Sea-Grass Mortality Along a Low-Energy Barrier Island Coast: West-Central Coast." *Journal of Sedimentary Petrology*, vol. 57, pp. 431-439, 1987.

Stapor, F.W., T.D. Mathews, and F.E. Lindfors-Kearns. "Episodic Barrier Growth in Southwest Florida: A Response to Fluctuating Holocene Sea Level." *Memoir 3*, pp. 149-202. Miami Geological Society, 1988.

Figure Credits

All figures not otherwise noted are by the author.

Figure 1.1 - Google Earth

Figure 2.10 - Brad Tritschler

Figure 2.28a – Google Earth

Figure 2.29 – Google Earth

Figure 3.1 - Google Earth

Figure 3.5 - Albert C. Hine

Figure 3.7 - Bay Soundings

Figure 3.14 - Google Earth

Figure 3.31 - U.S. Geological Survey

Figure 3.35 - MessyMessyChic

Figure 4.13 - Florida Vacation Rental

Figure 4.15b - David MeRee

Figure 4.16 - Google Earth

Figure 4.20 - Google Earth

Figure 5.6 - Lifestyles

Figure 5.8 - Sussex UK

Figure 5.31 - Google Earth

Figure 6.21 - Google Earth

Figure 7.12 - Google Earth

Figure 7.18 – Sanibel Captiva Fishing

Figure 7.19 - Flickr

Figure 7.23 - Humiston and Moore

Figure 8.6 - Google Earth

Figure 8.12 - Google Earth

Figure 8.13 - Island Real Estate

Figure 8.14b - University of Florida Archives

Figure 8.18 - Google Earth

Figure 8.19 - Frank Stapor

Figure 8.25 - David McRee

Figure 9.11 - Google Earth

Figure 9.12 - USF Archives

Figure 9.13 - Zack Tyler

Figure 9.15 - Jim Pike

Figure 9.16 - Zack Tyler

Figure 9.17 - Zack Tyler

Figure 9.18 - Zack Tyler

Figure 9.21 - Google Earth

Figure 9.22 - Frank Stapor

Figure 9.27 - Collier County

Figure 9.28 - Alvin Lederer

Figure 9.29 - Marco Island Historical Society

Figure 9.30 - Marco Island Historical Society

Figure 9.31 - Marco Island Historical Society

Figure 9.32 - Alvin Lederer

Figure C.1 – Florida Memory Archives

Glossary

accretional – The adding on of material; generally sand to the barrier.

backbarrier – The backside or landward side of a barrier.

backbarrier estuary – The water body that lies landward of the barrier island and separates it from the mainland.

backbarrier wetlands – The vegetation community on the landward side of the barrier that is dominated by mangroves, typically reds, and may have some salt marsh plants.

bathymetry – Depth of water; mapping the floor of a water body.

beachrock – Beach material that was lithified in the beach environment.

borrow material – Material taken from a source of some location to nourish the beach.

catseye pond – A narrow water body between adjacent beach-dune ridges.

continental shelf – The most landward major province of the continental margin of the marine basins. It extends from the shoreline to the break in slope at the continental shelf. On the Florida margin the shelf is wide, gently sloping, and mostly limestone with a thin veneer of mostly shelly sediment.

coppice mounds – Small incipient sand dunes.

downdrift end – The end of a barrier that is receives longshore sediment transport.

dredge spoil – Material taken by a dredge.

dredge-and-fill – Taking material with a dredge and filling adjacent locations.

drumstick barrier island – A mixed-energy barrier island that has a wide end due to accretion and a narrow end where washovers are common.

ebb shoal – The sediment body that accumulates on the open water side of a tidal inlet.

fillet – The accumulation of sand at a barrier structure such as a jetty.

flood shoal – The sediment body that accumulates on the landward end of a tidal inlet.

geotextile – A type of synthetic material

geotextile groin – A groin made out of geotextile tubes.

groin – A hard structure of various materials that is typically perpendicular to the shoreline and is designed to hold sediment.

Holocene – The most recent portion of the geologic time scale. It is designated to be from 10,000 years ago to present and all fossils are older than that. All present barrier islands on the Gulf of Mexico developed within this time frame.

hotspot – A local area where erosion of a nourishment or natural beach is abnormally high. The can be due to unusual wave refraction coupled with an abnormal topographic pattern.

intertidal – A position that is between low tide and high tide in elevation.

littoral – The shallow water against the shoreline.

longshore current – A current that is caused by the approach and refraction (bending) of waves to the shoreline. This current moves parallel to the shoreline and can transport sediment.

longshore sediment (or sand) transport – The movement of sediment along the shoreline produced by wave-generated currents.

mangal – The term used to refer to a stand of mangroves.

mediterranean – A marine basin that is smaller than an ocean such as the Gulf of Mexico and the Mediterranean Sea.

Miocene – An epoch in the Tertiary Period on the geologic time-scale that is 5-23 million years before present.

mixed-energy barrier island – Barrier island that is wide at one end showing sediment accumulation and progradation but narrow at the other end indicating sediment starved conditions. The drumstick shape is typical of these sediment bodies.

morphodynamics - The process/response system that takes place along the coast or in any other sedimentary environment.

morphology – The shape of things.

nourishment– In this context it is the addition of sand to the beach in order to widen it.

Pleistocene – The most recent epoch in the geologic time scale. It includes the last 2.6 million years during which glaciers were widespread throughout the high latitudes of the world.

prograde – Moving forward.

recurved – Making a corner.

recurved spit – An extension of a barrier island that is commonly influenced by tidal flux and becomes curved or bent landward.

relict – Old and not now active.

renourishment – To nourish again.

revetment – A built accumulation of large rock pieces placed to protect the shoreline.

ridge and runnel – A typically intertidal sand bar and adjacent trough that result from storm waves removing sand from the beach and placing it just seaward.

rip-rap – Large rock debris used to protect the shoreline.

semi-diurnal tidal cycle – Two tidal cycles (flood and ebb) of 12 hours and 25 minutes per tidal day.

shell hash – Hash not of corned beef but of shell debris.

shoaling – Becoming shallow.

shore-normal – Perpendicular to the shore.

spit – A linear extension of the sandy shoreline zone.

storm surge – Elevated water level caused by friction between the air and water during strong onshore wind. On this coast it is regularly more than the astronomical tide.

stratigraphic – Refers to the layering of sediment into vertical sequences

subtidal – Below low tide

supratidal – The environment just above high tide that is not flooded except during storm tides.

swash bar – Sand bars that are typically intertidal and that are moved landward by wave-generated motion.

swash zone – Typically equivalent to the foreshore portion of the beach, it is where the waves rush up and back on that surface.

tectonic – Refers to movements of the earth's crust

terminal groin – A structure placed at the end of a barrier or a portion of a barrier to hold the beach sand in place.

terrigenous – Having a land origin. This refers to sediment derived from erosion of rock bodies on land.

tidal flux – The rate of movement of water that is exchanged during a tidal cycle.

tidal inlet – Channel between barrier islands through which tidal currents flow

tidal prism – The volume of water that passes through a tidal inlet during each tidal cycle.

tidal range – The vertical amount between the low tide elevation and high tide elevation.

tide-dominated coast – A coast where the dominant processes is tidal flux. It does not develop barrier islands.

trailing edge coast – The coast on a land mass that is moving away from an oceanic ridge such as the east coast of North and South America.

uniformitarianism – A fundamental axiom of geology that says "the present is the key to the past"

washover fan – A generally thin, fan-shaped sediment accumulation produced by high water and waves. This sediment accumulation is generally the site of the wetlands on the landward side of the barriers.

wave-dominated barrier island – Long, narrow barrier islands where the primary energy to leads to formation is produced directly or indirectly by waves.

wave refraction – The bending of waves as they move through shallow water and are influenced by water depth.

Index

Dr. Richard A. Davis Jr. was a professor of geology at the University of South Florida until 2005. He is now an emeritus distinguished university professor at USF and visiting professor/research associate at Harte Research Institute, Texas A&M University in Corpus Christi, Texas. His area of specialization is coastal geology with emphasis on beaches, barrier islands, and tidal inlets. He has written/edited 20 books and about 150 journal articles on topics of oceanography, coastal geology, stratigraphy, and sedimentology. He has been a visiting professor in Australia, New Zealand, the Netherlands, Denmark, Spain, and Germany, and he speaks and teaches workshops at universities all over the world. He holds the Shepard Medal in Marine Geology.

His previous books include *Beaches in Space and Time: A Global Look at the Beach Environment and How We Use It,* published by Pineapple Press, which offers a look at the beaches of the world, how they have evolved to their present form. He also has recently published *Beaches of the Gulf Coast* and *Sea-Level Change in the Gulf of Mexico,* both in the series from Harte Research Institute for Gulf of Mexico Studies.

Praise for Beaches in Space and Time:
"This book by an internationally well-known coastal expert translates science into comprehensible popular language. Readers will especially appreciate the profusion and variety of illustrations from the entire globe."

— Dr. Ervin G. Otvos, *Head Geology Section, Gulf Coast Research Laboratory and Emeritus Professor, Department of Coastal Sciences, University of Southern Mississippi*

Here are some other books from Pineapple Press on related topics. For a complete catalog, write to Pineapple Press, P.O. Box 3889, Sarasota, Florida 34230-3889, or call (800) 746-3275. Or visit our website at www.pineapplepress.com.

Beaches in Space and Time by Richard A. Davis Jr. This is a global look at the beach environment and how we use it. Hundreds of color photos, charts, and diagrams.

Florida's Living Beaches by Blair and Dawn Witherington. Detailed accounts of over 800 species, with color photos for each, found on Florida's sandy beaches. Covers plants, animals, minerals, and manmade objects.

Florida's Seashells by Blair and Dawn Witherington. Accounts, maps, and color photos for over 250 species of mollusk shells found on Florida's beaches

Living Beaches of Georgia and the Carolinas by Blair and Dawn Witherington. Lists over 850 items found along 600 miles of Atlantic coastline. Color photos highlight birds, turtles, fish, mammals, flowers, and much more.

Seashells of Georgia and the Carolinas by Blair and Dawn Witherington. Color photos of hundreds of the shells you'll find on the beaches of Georgia and the Carolinas, including details about the features, habitats, and diet of each shell's inhabitant.

Our Sea Turtles by Blair and Dawn Witherington. A comprehensive narrative of every aspect of sea turtles' lives, from egg laying to human rescue efforts. Includes stunning color photos, maps, illustrations, and charts that reflect sea turtles' unique contributions to our environment. Meticulously researched.

Best Beach Games by Barry Coleman. Discover 75 simple, engaging games you can play on the shore with your kids. Most of the games require nothing more than items normally found on the beach.

Just Yesterday on the Outer Banks, Second Edition, by Bruce Roberts and David Stick. Celebrates the history and uniqueness of North Carolina's Outer Banks with photos and text that feature shipwrecks, lighthouses, fishermen, beaches, and even picket fences. Historian David Stick's gentle prose accompanies images by renowned photographer Bruce Roberts.

Shipwrecks, Disasters & Rescues of the Graveyard of the Atlantic and Cape Fear, Second Edition, by Norma Elizabeth and Bruce Roberts. Covering 1750 to 1942, this slim volume highlights the most famous shipwrecks and sea disasters that occurred off the coast of North Carolina. Filled with color and black-and-white historical illustrations and contemporary photographs, this is a treasure trove of facts and details about these wrecks and rescues.

Shipwrecks of Florida by Steven Singer. The most comprehensive listing now available of over 2,100 shipwrecks from the 16th century to the present. Extensive appendices offer a wealth of information for divers and researchers.

Bansemer's Book of the Southern Shores by Roger Bansemer. An artist's journal describing in words and paintings the natural beauty of the coasts—from the sponge divers of Tarpon Springs to the marshlands of coastal Georgia.

Guardians of the Lights by Elinor DeWire. Stories of the fortitude and heroism of the men and women of the U.S. Lighthouse Service, who kept vital shipping lanes safe from 1716 until early in the 20th century.

The Lightkeepers' Menagerie by Elinor DeWire. Stories of animals that have lived at lighthouses, including bell-ringing dogs, swimming cats, parrots, deer, bears, foxes, horses, mules, goats, and cows. This thick volume is loaded with vintage photos and endearing stories of animal companions and includes an 8-page color section.

The Lighthouses of Greece by Elinor DeWire and Dolores Reyes-Pergioudakis. Lavishly illustrated and carefully researched, this full-color book covers over 100 lighthouses, most still guiding ships around the Greek Islands.

Lighthouse Families, Second Edition, by Bruce Roberts and Cheryl Shelton-Roberts. What was it like to live and work at a lighthouse during the heyday of shipping and fishing? Filled with first-person accounts and loads of family photos, this is a record of the memories and stories of America's lighthouse keepers.